Stratification and Inequality Series
The Center for the Study of Social Stratification and Inequality,
Tohoku University, Japan
Volume 3

I0129255

Constructing Civil Society in Japan

Stratification and Inequality Series

The Center for the Study of Social Stratification and Inequality,
Tohoku University, Japan
Volume 3

Inequality amid Affluence: Social Stratification in Japan
Junsuke Hara and Kazuo Seiyama

Intentional Social Change: A Rational Choice Theory
Yoshimichi Sato

Constructing Civil Society in Japan: Voices of Environmental Movements
Koichi Hasegawa

Stratification and Inequality Series

The Center for the Study of Social Stratification and Inequality,
Tohoku University, Japan
Volume 3

Constructing Civil Society in Japan

Voices of Environmental Movements

Koichi Hasegawa

First published in 2004 by
Trans Pacific Press, PO Box 164, Balwyn North, Victoria 3104, Australia
Telephone: +61 (0)3 9859 1112 Fax: +61 (0)3 9859 4110
Email: tpp.mail@gmail.com
Web: http://www.transpacificpress.com

Copyright © Trans Pacific Press 2004

Designed and set by Digital Environs, Melbourne, Australia. www.digitalenvirons.com

Distributors

Australia and New Zealand
DA Information Services/Central Book
Services
648 Whitehorse Road
Mitcham, Victoria 3132
Australia
Telephone: +61-3-9210-7777
Fax: + 61-3-9210-7788
Email: books@dadirect.com
Web: www.dadirect.com

USA and Canada
International Specialized Book
Services (ISBS)
920 NE 58th Avenue, Suite 300
Portland, Oregon 97213-3786
USA
Telephone: 1-800-944-6190
Fax: 1-503-280-8832
Email: orders@isbs.com
Web: http://www.isbs.com

Asia and the Pacific
Kinokuniya Company Ltd.

Head office:
3-7-10 Shimomeguro
Meguro-ku
Tokyo 153-8504
Japan
Telephone: +81-3-6910-0531
Fax: +81-3-6420-1362
Email: bkimp@kinokuniya.co.jp
Web: www.kinokuniya.co.jp

Asia-Pacific office:
Kinokuniya Book Stores of Singapore Pte., Ltd.
391B Orchard Road #13-06/07/08
Ngee Ann City Tower B
Singapore 238874
Telephone: +65-6276-5558
Fax: +65-6276-5570
Email: SSO@kinokuniya.co.jp

ISBN   978-1-876843-67-0 (Hardback)
ISBN   978-1-876843-73-1 (Paperback)

National Library of Australia Cataloging in Publication Data

Hasegawa, Koichi, 1954–.
    Constructing civil society in Japan : voices of
    environmental movements.

    Bibliography.
    Includes index.

    ISBN 978-1-876843-67-0 (hbk).
    ISBN 978-1-876843-73-1 (pbk).

    1. Environment protection – Japan. 2. Environmental policy
    – Japan. 3. Japan – Environmental conditions. I. Title.
    (Series : Stratification and Inequality series).

363.70560952

Contents

List of tables

List of figures

To the late Professor Nobuko Iijima (1938–2001),
a true pioneer and the
'Mother of Environmental Sociology'

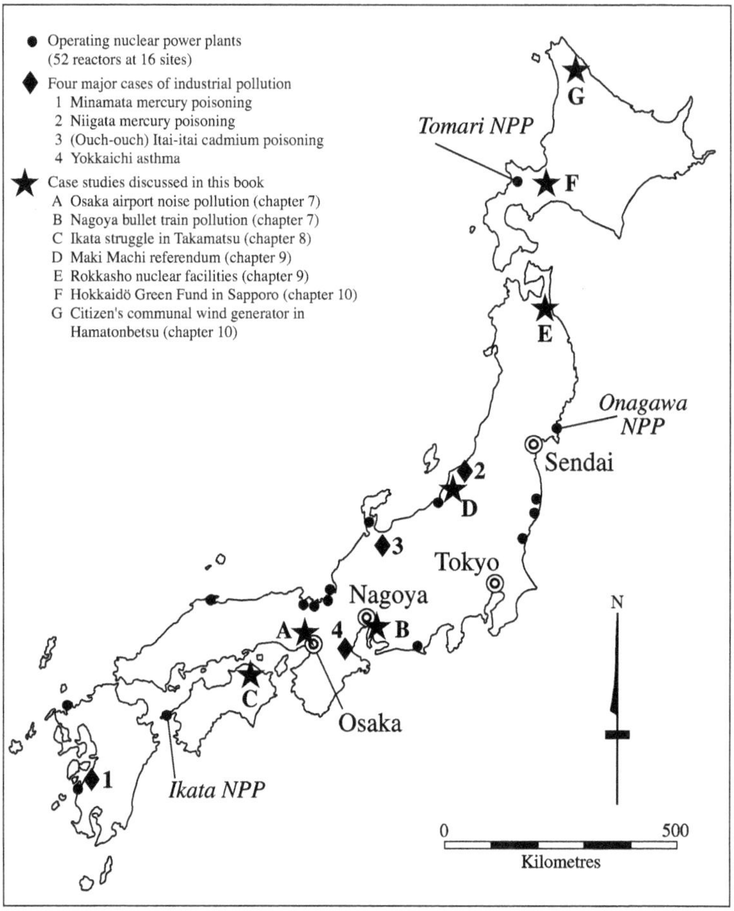
Operating nuclear power plants
(52 reactors at 16 sites)
Four major cases of industrial pollution
1 Minamata mercury poisoning
2 Niigata mercury poisoning
3 (Ouch-ouch) Itai-itai cadmium poisoning
4 Yokkaichi asthma
Case studies discussed in this book
A Osaka airport noise pollution (chapter 7)
B Nagoya bullet train pollution (chapter 7)
C Ikata struggle in Takamatsu (chapter 8)
D Maki Machi referendum (chapter 9)
E Rokkasho nuclear facilities (chapter 9)
F Hokkaidō Green Fund in Sapporo (chapter 10)
G Citizen's communal wind generator in Hamatonbetsu (chapter 10)
Tomari NPP
Onagawa NPP
Sendai
Tokyo
Nagoya
Osaka
Ikata NPP
N
0
500
Kilometres

Figure 0.1: Environmental issues in Japan discussed in this book

Preface to the English edition

To date, the Japanese social sciences have generally demonstrated strong tendencies to debate earlier theories and to introduce the latest theoretical developments from the West. Efforts to face and tackle the concrete social problems and policy issues confronting Japanese society and to develop theories based on primary research remain weak. In this situation environmental sociology is a field of study that strongly emphasizes original contributions. The term 'environmental sociology' was born in the USA in the mid 1970s. In Japan, Nobuko Iijima, who later became the first president of the Japanese Association for Environmental Sociology, began conducting sociological studies of environmental pollution issues in the late 1960s. Her paper, 'Industrial pollution and the local residents' movements (Sangyō kōgai to jūmin undō)', published in 1970 (Iijima, 1970b), compared the Minamata mercury poisoning disease to the Niigata Minamata disease. As well as revealing the multiplicity and multi-layered nature of the damage, the paper analyzes how the damage and the response from general residents are defined by the local community. Internationally, this can be regarded as the first paper in authentic environmental social science research.

Hiroyuki Torigoe and others wrote 'Environmental history of water and people (*Mizu to hito no kankyō-shi*)' (1984), a distinguished study unique to Japan, utilizing folklore to reveal how the local culture of the people around Lake Biwa was intertwined with their relationship to the water and water management system. They called their theoretical and methodological paradigm 'life-environmentalism'.

The joint research of Harutoshi Funabashi and myself, published as 'Bullet train pollution (Shinkansen Kōgai)' (1985), was the world's first dedicated sociological study of traffic pollution. It revealed social factors that contributed to the noise and vibration pollution arising in the shadow of Japan's world-class bullet train, the social mechanisms that made the issues difficult to resolve, and the social forces that carried forward the measures to rectify the bullet train noise and vibrations (see chapter 7).

There have been many other pioneering studies in Japanese environmental sociology (chapters 1 and 5). However, most of these were written in Japanese, and very few English editions have yet been published. Thus, except for those few environmental sociologists who have personal contacts abroad, the significance of this research has not been sufficiently recognized by the international academic community.

This book is written in the tradition of Japanese environmental sociology, with an emphasis on fieldwork and case studies. It also focuses on the new social movements and trends in Japan after the late 1980s.

In the 1950s, '60s, and '70s Japan enjoyed levels of economic prosperity that were referred to as the 'Japanese miracle', the 'dark-side' of which included widespread and serious pollution problems, such as the four major industrial pollution cases (beginning with the Minamata disease mentioned above). The vigor of Japan's independent environmental research is a reflection of the severity of the pollution, as well as the political and social pressures from industry and government for large-scale industrial development.

The so-called economic miracle came to an end, followed by a deadlock in political and economic reform that has been referred to as the 'lost 10 years'. Since the late 1980s, however, there have been new movements to vitalize Japan's civil society (see chapters 8, 9, 10, 13 and the Conclusion). Especially since the mid '90s, this can be seen in the sudden increase of certified incorporated NPOs, the increasing number of actions seeking the disclosure of public information, local referendums and policy-oriented social movements.

The policy-making process for Japanese environmental policies remains relatively closed by international standards. With opportunities such as the 1992 Earth Summit, a 'new public sphere' concerned with environmental issues and policies was also born in Japan, and gradually opened to its citizens. This book is a sociological review of the environmental movements in contemporary Japan, as well as the 'new public sphere', the vibrant civil society that the movement supports.

The 'public sphere' is a place for the formation of public opinion and social agreement. This is where people with public interests gather, to discuss what the 'public interest' is, to carry out social practices, to realize 'publicness' and 'communality', and to carry out political education. As Bellah et al. explain: 'Here citizens develop new hopes through the practice of public conversation and joint action' (Bellah et al, 1991:269). This normative public sphere that is slowly replacing the traditional, closed Japanese public sphere, is called the 'new public sphere' in this book.

The stereotypical image of the obedient Japanese who silently follows the traditional order is still strong internationally, but the voices of people interested in environmental movements and environmental NGOs in the new public sphere are loud and diverse. The voice of the environmental movement calls for a new wind, and this wind in turn offers a powerful, sound energy to civil society, just as electricity is produced by power-generating windmills.

In contrast to environmental economics or environmental law, the unique perspective of environmental sociology lies in its analysis of environmental movements. In many countries around the world, including Japan, the USA, and European nations, the environmental movement is the 'father' of environmental sociology. This book, using the perspective of environmental sociology, analyzes the structure, the dynamics and the task of the Japanese environmental movement, which started as an anti-pollution movement and local residents' movement.

This book also aims to be a portrait of modern society, focusing on the dynamics of the environmental movement and the public sphere. The environmental movement is a realm of 'exemplary action', as well as a 'trail-blazer'. It has shone new light on the needs of the citizens and, through collaboration with the government and industry, has delivered numerous pioneering efforts. If the 20th century was the century of economic growth and industrial capital, countless people around the world hope that the 21st century will be the century of the environment and the citizens. To this end, the environmental movement and the new public sphere it involves may become a compass for modern society.

The possibilities for the civil societies of East Asia have received much international attention with the progress of democracy in South Korea and Taiwan in the 1990s, and the boom in NGOs. The possibility of a new civil society developing in China, especially, will have a major impact on regional and global politics in the first half of the 21st century. The voice of Japan's civil society may strengthen similar voices in East Asia and generate a new wind for the environmental and citizens' movements.

This book was originally published under the title *Kankyö undö to atarashii kökyöken—kankyö shakaigaku no päsupekuchibu* (Environmental movements and the new public sphere—the perspective of environmental sociology), by Yūhikaku, in April 2003.

It is a great pleasure that the English edition has been produced in such a short time (just over one year). This is largely due to the care and encouragement of Professor Yoshio Sugimoto, School of Social Sciences, La Trobe University and Director of Trans Pacific Press, who understood the significance of this volume. I thank him sincerely. I must also thank the team of English translators: David Askew, Anna Dobrovolskaia, Jun Nakagawa, Esther Rockett, Ania Siwicki and Rick Tanaka. My conversation with the TPP editor, Karl Smith, who provided detailed comments on each chapter, was a most stimulating contribution to the preparation of this English edition.

Professor Jeffrey Broadbent of the University of Minnesota has given me support since 1988, and invited me into the international academic network of environmental sociologists and social movement researchers. It is due to his assistance that I am able to stay in the Netherlands and the USA during 2004–5, as an Abe Fellow of the US Social Science Research Council and as a visiting professor at the University of Minnesota. Without his enduring encouragement and friendship, the publication of this book would not have been possible.

I am very happy and honored to have the chance to publish this book as part of the CSSI (Center for the Study of Social Stratification and Inequality, Tōhoku University) series by Trans Pacific Press. This is the third volume of the series and the result of the CSSI's research in the 21st Century Center of Excellence Program formed by the Japanese Ministry of Education, Culture, Sports, Science, and Technology.

Koichi Hasegawa
May 2004

Bara no me ya, kōsei oeshi eibunkō

Rosebuds opening.
Just finished proofing
English manuscript

Part I:
The principles and issues of environmental sociology

1 Perspectives of environmental sociology: The issues of the second stage

From systemization to the second stage

The systemization of environmental sociology

Environmental issues, the state of the environment and civilization, and the relationship between the environment and human beings have become important issues for research in the social sciences in recent decades. Interest in environmental issues is increasing in the field of sociology. During the last decade of the twentieth century, environmental sociology was established in Japan as a concept and a field of deliberate research analysis. Environmental sociology has become socially recognized and institutionalized, at least academically. Alongside specializations such as gender studies, historical sociology and ethnic studies, environmental sociology has in recent years been one of the most active fields of sociological research, both in Japan and overseas.

The level of research activity is evident in the number of new academic organizations, their growing membership and increasing numbers of publications. The JAES (Japanese Association for Environmental Sociology) began on 19 May 1990 as a small group of environmental sociologists with 53 members. It changed its title to JAES and formally organized in October 1992. As of 28 June 2003 it had 657 members. Figure 1.1 charts the membership increase. In twelve years the membership has increased eleven-fold, with a net increase of more than fifty new members per year, and the JAES has begun nurturing young researchers. The JAES is the largest organization specializing in environmental sociology in the world. It is also one of the largest academic associations in the field of sociology in Japan, alongside the Japanese Association for Family Sociology.

In September 1995, the annual official *Journal of environmental sociology* (*Kankyō shakaigaku kenkyū*) was founded. It is the world's oldest specialist journal in the field of environmental

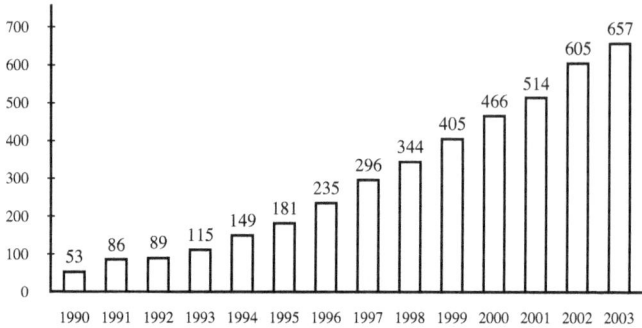

Figure 1.1: Membership of the Japanese Association for Environmental Sociology (1990–2003)

sociology, followed by the Korean Association for Environmental
Sociology's journal ECO, founded in 2001. The Korean Association
for Environmental Sociology was founded in 2000, growing out of
an informal group set up in 1995. Its organizational structure and
activities are modeled on the JAES.

One key characteristic of environmental sociology is that many
researchers in the field are from areas outside a narrowly defined
field of sociology, and some are practitioners of environmental
movements, government administrations and private firms. Only a
third or so of the members are also members of the Japan Sociological
Society, the professional association of academics and post-graduate
students specializing in sociology. The other 70% are from the
natural sciences (such as agriculture) or other social sciences (such
as economics, management and law), and non-researchers such as
NGO activists.[1] In comparison, specialist organizations in sociology,
such as the Japan Association for Urban Sociology and the Japan
Association of Regional and Community Studies typically have
about 300 members. The JAES's membership continues to attract
members from activists and other research areas, although it did
not intentionally aim to expand its membership from these areas.
Environmental issues are inherently interdisciplinary in character,
and therefore the expectations of environmental sociology from
'outside' of sociology were high.

Despite the increasing difficulty in publishing academic work
in Japan, a systematic five volume monograph series entitled
Environmental sociology in Japan (*Kōza kankyō shakaigaku*) was
published by Yūhikaku in 2001. This series had 44 contributors,

of whom 43 were members of the Association of Environmental Sociology. A more introductory six volume set entitled *Environmental sociology: A series* (*Shirīzu kankyō shakaigaku*) was published by Shinyōsha between 2001 and 2003. Nobuko Iijima, who passed away in 2001, was a true pioneer in this field at an international standard and was honored with the name 'Mother of environmental sociology'. She began her sociological research on environmental issues in the late 1960s, researching the victims' life structures caused by Minamata and Niigata mercury poisoning—two of the four major environmental problems (Iijima 1970b). With the exception of residents' movements that arose in the face of problems with large-scale development,[2] she struggled alone in Japan until the 1980s. In the past ten to fifteen years, however, there has been a dramatic change in the research environment for environmental sociology in Japan.

The USA does not have an independent academic association like Japan's Association for Environmental Sociology, although there is a section of the American Sociological Association which is focused on 'Environment and Technology' (it was originally called Environmental Sociology, but changed its name in 1988). Established in 1976, over the past decade its membership has remained stable at around 400.[3] Researchers from other areas only rarely participate, and almost all of its members are sociologists. During the Reagan administration (1980s), the membership gradually declined and the section stagnated. However, it began to revive around 1990 and its membership recovered, with a growing interest in global environmental issues.

At the global level, a Research Committee (RC 24) entitled 'Environment and Society' exists within the International Sociological Association (ISA). In its early days, most of its members were European researchers, but since the 1990s, participation by researchers from the USA has been increasing. It was founded as a Working Group with about 40 participants in 1990, and was promoted with unprecedented speed (for the ISA) to a Research Committee in 1994.[4] In 1998 there were about 150 members, and in 2002 there were about 200.[5] The first president of this committee was R. Dunlap (1994–98), an early proponent of 'environmental sociology'. He was followed by F. Buttel (1998–2002), perhaps best known for his debate with Dunlap (see below). Both Dunlap and Buttel were from the USA, but the third president, A. Mol (2002–), is from the Netherlands. Mol is a proponent of 'ecological modernization'. In short, although a relatively new field, the numbers of researchers with interests in environmental sociology are increasing in both Japan and elsewhere, and new organizations are still being founded. Until the early 1990s environmental research around the world was primarily conducted in the natural sciences. However, from the second half of the 1990s, environmental research has received increasing attention in the social sciences. In Japan, the Society for Environmental Economics and Policy Studies was founded in 1995, mainly dominated by economists. Beginning with a membership of 866, by September 2002 it had 1,339 members. The Japan Association for Environmental Law and Policy was founded in 1997. Dominated by legal scholars, it had 415 members as of May 2002. Both associations are active in conference activities as well as academic research. Together with the Association of Environmental Sociology, these two associations form the core membership of a multi-disciplinary symposium called the Frontiers of Environmental Policy Research, which has been held annually since 2000 (Awaji et al. 2001). Unfortunately, there are very few researchers in the fields of psychology, politics or administration studies who have an interest in environmental issues or environmental research, and there are no indications that associations of environmental research are being organized in these fields.

The social background to the emergence of environmental studies

The growing (global) public interest in environmental issues has come in two waves, both culminating in environmental conferences organized by the UN. The first wave was evident at about the time of the 1972 UN Human Environmental Conference held in Stockholm, while the second wave became apparent at about the time of the 1992 UN Environment and Development Conference (the Earth Summit) held in Rio de Janeiro.

The first wave is closely related to the Apollo moon landing in July 1969, the Earth Day of April 1970, and the publication of *The Limits of Growth* by the Rome Club in 1972 (Meadows et al. 1972). As the Apollo project unfolded to reveal the barrenness of the moon, there was a heightened global consciousness that the Earth is irreplaceable. At the same time an increasingly critical attitude towards advanced scientific technology developed around the world, driven by concerns about the high rates of economic growth and the corresponding waste of resources. The growth of environmental sociology in the

USA in the mid 1970s occurred against this background. In Japan, there was a special set of circumstances, with savage environmental problems such as the four major environmental problems, such as the Minamata mercury poisoning (see chapter 3 note 3) and a growing social disquiet about large-scale development. In 1970, the concept of environmental rights was proposed, and in 1971 a pioneering specialist journal was founded, entitled *Research on environmental disruption* (*Kankyö to kögai*, originally titled *Kögai kenkyü*). The second wave corresponds to the debate arising in the late 1980s about the problems of global warming. During the Earth Summit of 1992, Agenda 21 was put together, the UN Convention for Climate Change Framework was ratified, and an international framework of arrangements to address the global environmental issues of the 1990s was established. Broadly speaking, the organization and systematization of groups of social scientists specializing in environmental sociology and environmental policy studies was a response to these currents, both in Japan and overseas.

The main theme of environmental research in the natural sciences is to measure environmental costs and destruction, and to identify the causes and effects of the physical mechanisms involved. As other issues, such as the most effective policy response and how to implement it, the conflicting interests of the societies of the North and South, the relationship between industry and the citizenry, and changes in everyday behavior and environmental consciousness have become increasingly important, there has been an accompanying emphasis on social science research. As interest has moved from identifying causes and effects to policy studies, so too, have expectations for sociological and social scientific research grown.

Issues of the first stage

Having reached the twenty-first century, environmental sociology has also reached its first major turning point since it began as an academic field of study and its professional association was founded. In order to examine the strategic issues of the second stage of environmental sociology it is necessary to first assess the accomplishments of the first stage.

This section summarizes the first stage issues of environmental sociology. As mentioned, membership in the professional associations in this field has been steadily increasing. The association journal, and publications such as the aforementioned monograph series and six-volume set have contributed to growing social recognition of environmental sociology as an academic field of study. Since the Association of Environmental Sociology was founded, many joint research projects have been conducted, led by researchers such as Iijima.[6] The key issue of the first stage was perhaps to heighten the concentric and internal perspectives through the organization and systematization of an academic institution.

Among the first stage issues, the most contentious issue is the extent to which environmental sociology has established an academic identity. What kind of academic field of study is environmental sociology? It might be said that environmental sociology (1) researches the mutual interaction between human beings, society and the environment, and (2) in more concrete terms can be divided into 'the sociology of environmental problems' and 'the sociology of environmental coexistence'.[7] These questions were explicitly addressed in *Environmental sociology in Japan*.[8] The first volume attempts to provide an overview of the field of environmental sociology as an academic field of study. Volumes 2, 4 and 5 discuss concrete developments in 'the sociology of environmental problems', including discussions of perpetrators, victims, movements and policies. Volume 5 addresses these issues with a particular focus on Asia. Volume 3 explores 'the sociology of environmental symbiosis'. I was personally involved in planning and editing this series, so others should judge the extent to which its objectives were achieved. However, I believe that this series was more comprehensive than similar series in other academic fields. The continuing publication of edited textbooks (Iijima ed. 1993, Funabashi and Furukawa eds, 1999), single-author textbooks (Kada 2002), and monograph series suggests that environmental sociology has indeed established a mature academic identity.

Issues of the second stage

This section outlines the second stage issues of environmental sociology. The strategic direction of the second stage is centrifugal, as new developments are made possible through links outside of environmental sociology and its association and as the gaze of environmental sociologists turns and expands outwards.

(1) The key issues concerning the relationship with the 'research field' of environmental issues—primarily victims and sufferers, residents and citizens, environmental NGOs and activists,

and the authorities in charge of government policy—will be to advance and deepen policy studies while emphasizing its importance in environmental sociology.

(2) The main issues relating to sociological theory or 'mainstream sociology' will be to strengthen the bridges to core theoretical areas while deepening the theoretical aspects of environmental sociology itself.

(3) It is necessary to actively promote further interdisciplinary research and dialogue in relation to associated areas of environmental research, such as environmental law, economics, ethics, and medicine.

(4) An urgent task is for Japanese environmental sociologists to increase their international transmission of knowledge by strengthening relationships with overseas environmental sociology, related academic associations and researchers overseas.

Of the four sets of issues, the author places a special emphasis on environmental policy studies as an area where Japan's environmental sociology should make greater efforts in the future. The relationship between environmental sociology and related policy studies are discussed in detail in chapter 6. Here, the focus is on the issues of interdisciplinarization and international knowledge transmission. Building theoretical bridges with mainstream sociology will be examined in the next section.

The significance of and issues in interdisciplinary joint research

As environmental sociology further deepens its research interaction with other disciplines—such as environmental economics, policy studies and law—public policies relating to environmental problems will be the key issues of common concern. To date, there has been very little research in Japan on environmental problems in the fields of political science and public administration studies, with the exception of a handful of researchers such as Yorimoto Katsumi. Why researchers in political science and public administration studies have been so uninterested in environmental policies and environmental research is a question that requires examination in and of itself.[9]

What are the characteristics and issues of Japan's environmental policies and policy-making processes when viewed in the international context? One of the most crucial tasks environmental

Increase membership

Organizing

Establish an
academic identity

Social recognition

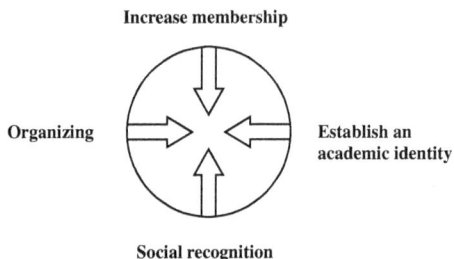

*Figure 1.2: Focus of the first stage of Japanese enviromental
sociology*

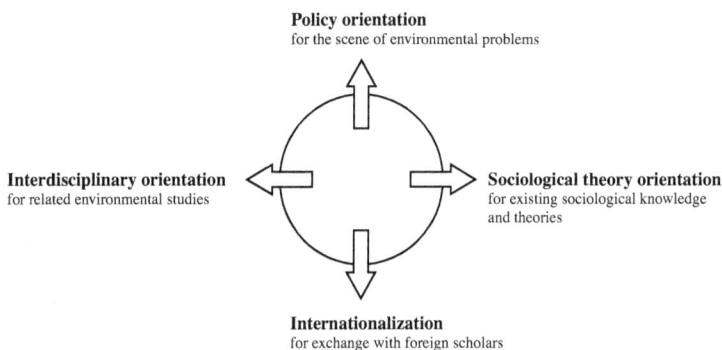

Policy orientation
for the scene of environmental problems

Interdisciplinary orientation
for related environmental studies

Sociological theory orientation
for existing sociological knowledge
and theories

Internationalization
for exchange with foreign scholars

*Figure 1.3: Focus of the second stage of Japanese enviromental
sociology*

research has to tackle is to shed light on this question. For instance, countries such as Germany, Denmark and the Netherlands are particularly active in developing policies to tackle global warming. Why these particular countries are so enthusiastic about developing such policies in relation to the novel characteristics of their social structures and decision-making processes is a new question that environmental sociology should come to grips with, using comparative political systems theory and comparative sociology and working with researchers in other societies and areas of environmental research.[10]

Since the late 1980s, there has been a large shift in environmental and energy policies in the USA and northern European countries as a result of collaboration between environmental NGOs and environmental NPOs, government agencies and enterprises. This provides a stark contrast to the closed nature of Japan's policy-making processes under what has been called 'structured paternalism' (Yonemoto 1994: 229). These differences highlight the importance of collaboration between environmental sociology, economics, and policy studies with the authorities in charge of policy, as well as environmental NGOs and NPOs for researching and formulating environmental policies.

Joint research with scholars of environmental economics and policy studies is essential for fostering interdisciplinary interaction, but to date there has been little joint research including environmental sociologists. It is thus necessary to continue to refine the specific tools of environmental sociology in order to ensure that joint research may be fruitful.

The necessity of international knowledge transmission

In both the pre- and postwar eras, sociology in Japan has shown a strong tendency to adopt the academic theories and research trends of foreign countries, i.e. to 'import' knowledge, without also trying to export it or to otherwise influence overseas developments through dialogue. In this regard, environmental sociology lags behind psychology and cultural anthropology as well as economics and policy studies. Since 1994, I have participated in the World Congress of the International Sociological Association held every four years, but in spite of the fact that it boasts the largest membership of any environmental sociology organization in the world, attendance and presentations by Japanese researchers in the field have been quite limited.[11] There have also been very few articles published in English that might be cited by international researchers. This is true of Japanese sociology in general, with the exception of social stratification and mobility research.

The paucity of Japanese sociological contributions to inter-national dialogue is in direct contrast to the unique and significant theoretical contributions of Japanese environmental sociology, such as perspectives on the social structures of victims (Iijima 1984/1993), the theory of benefit versus victimized zone (Funabashi et al. 1985), and life-environmentalism (Torigoe and Kada eds, 1984). Moreover, research has been conducted on topics barely touched by overseas researchers in environmental sociology, such as pollution from high-speed transportation, the preservation of historical streets and houses, the promotion of organic agriculture, and cultural attitudes towards water. With their knowledge of Chinese characters, researchers in Korea and China are able to read Japanese articles in the original, or in translation. However, because so little is published in English, Japanese works are rarely referred to by other international researchers. Nor is Japanese work widely known to Asian researchers who cannot read Japanese. It is therefore urgent that Japan's experiences and unique knowledge is internationally transmitted, to enrich theoretical perspectives and research frameworks through mutual interaction.[12]

The problem of environmental sociology's identity

Examining the theoretical bridges with mainstream sociology is essential to understanding environmental sociology as a unique academic field, and to considerations about the future of sociology itself. This kind of examination has only just begun both overseas and in Japan (Dunlap et al. eds, 2002 and Umino 2001).

A textbook explanation of environmental sociology might describe it as a field of academic study that sociologically analyzes the mutual interaction between environment and society (Humphrey and Buttel 1982). More simply, environmental sociology conducts sociological research about the environment; it has one foot in environmental research and the other in sociology. However, it would be too simplistic to define environmental sociology as a field of academic study in which sociological methodologies and approaches are applied to environmental issues and environmental consciousness.

The debate about the character of environmental sociology

The debate about the character of environmental sociology began in the USA during the second half of the 1970s, shortly after the field was first proposed. Dunlap and other advocates of environmental sociology argued that it was not merely a sociological analysis of environmental issues, but should also inform (and thereby reform) mainstream sociology's anthropocentric bias through its new ecological paradigm (referred to at the time as the 'new environmentalism' paradigm). Buttel and others criticized Dunlap for 'empty posturing', claiming that this 'new paradigm' had produced no notable results, and that environmental sociology ought therefore to persist with a sociological analysis of environmental issues (Buttel 1987). The author basically agrees with Buttel's criticism. Although the new value premises and perspectives proposed by Dunlap are attractive they fail to offer any new analytical tools or methodologies specific to environmental sociology. In other words, there has (as yet) been no 'paradigm shift' in Thomas Kuhn's sense of the term. Indeed, although Dunlap's most notable research was an international comparison of environmental consciousness, this was merely a particular variation of previously established research on social consciousness (see chapter 2 note 5).

Between environmental sociology and sociology

At the level of the individual researcher there is a growing potential for a distancing distinction between one's professional identity as an 'environmental sociologist' versus a 'sociologist', especially with regard to academic background and methodology (in terms of both education and supervision). As previously mentioned, about 70% of the 600 plus members of the Japanese Association for Environmental Sociology are not members of the Japan Sociological Society, but instead have backgrounds in other areas of research. Moreover, like the author, many environmental sociologists over the age of 40 were originally trained as sociologists and later developed an interest in environmental issues and related research, making this the major focus of their research. This group tends to identify themselves as sociologists first and foremost, who have moved towards a specialist identity as environmental sociologists. In contrast, younger researchers who began their academic careers with studies in the growing specialty field of environmental sociology or another discipline generally have limited understandings of basic sociological methods and theories except to the extent that these directly relate to environmental sociology. Moreover, environmental sociology is by its very nature both inter- and cross-disciplinary. Thus when conducting joint research or nurturing younger re-searchers, a major issue arises in, for instance, evaluating articles or conference papers: to what degree can the intangible understanding that 'this is sociological' (and 'that is not'), which has been internalized by the generation with sociological backgrounds be analytically objectified and communicated to younger researchers, and researchers from related areas?

This problem is exacerbated by a more general issue: as in academia more generally, sociology has tended towards increasing specialization in recent years. As with other (and more well-established) specialist sociologies, such as urban sociology[13] and family sociology, the distance between environmental sociology and the theoretical research that forms the core of mainstream sociology is relatively large. Here too there is the danger that the identity of researchers as 'environmental sociologists' and as 'sociologists' may become increasingly estranged.

Needless to say, specialist sociologies such as urban sociology, rural sociology and family sociology are not merely products of the application of the insights of sociological theories and methodologies to cities, farming villages and families. Each specialist sociology has generated unique theoretical and empirical perspectives, and has shown strong tendencies towards autonomy. In contrast to the other specializations mentioned, though, the distance between environmental sociology and the core areas of sociology is even larger due to the complete absence of discussions about the natural environment or environmental issues by the giants of sociology such as Durkheim, Weber and Parsons,[14] and the limited consciousness of environmental sociological problems in the sociological literature.

Before the field of environmental sociology was proposed, the 'proper' focus of sociology was long considered to be limited to the social environment, social relations, and the forms or structures of social groups, while the natural environment laid outside of this field. In Japan and the USA, environmental sociology did not emerge from the mainstream of sociology, but from peripheral areas such as rural sociology, regional and community studies, and social movement studies. This situations leads Umino Michio (2001) to the conclusion that the extent to which mainstream sociology and environmental sociology influence or are influenced by one another is limited. However, Dunlap et al. (eds, 2000) argue that the relationship between the two is deeper than the established sociological theories suggest. A further, more detailed and careful examination of the relationship between this sub-discipline and its 'parent' is required. In Japan at least, there are many points of contact between environmental sociology and rural sociology, regional and community studies, and the theory of social movements. Moreover, as discussed below, the direction of environmental sociology has been largely determined by the focus of researchers on various sociological variables and structural conditions.

The question of what environmental sociology can (and indeed *should*) learn from mainstream sociology, and to what extent, is directly related to questions of whether greater emphasis should be placed on its relationship with the educational and research activities of mainstream sociology, or its relationships with related academic fields of environmental research in an attempt to make it a sociological version of environmental studies.

Another important issue is the extent to which the development of environmental sociology can lead to a reformation of existing sociology as advocated by Dunlap and others. Indeed, the ecological modernization perspective of Mol and others (Mol and Spaargaren 2000) has had a significant impact on the theories of reflexive modernization developed by Beck, Giddens and Lash (1994). From this perspective, speaking metaphorically, we might say that if Marx's *Capital* were to be written today, the central struggle would be around environmental issues and their strategic importance in modern society.[15]

The danger of an identity crisis

As the distance between environmental sociology and mainstream sociology grows, so does the risk that environmental sociology will experience an identity crisis. Although the explosion in membership suggests social expectations for environmental sociology, it might alternatively indicate a belief among practitioners in related fields that environmental sociology is a 'soft' science, i.e. not highly specialized and therefore relatively accessible. In contrast, there has been no comparable growth in the membership of the Japan Association for Environmental Law and Policy.

If this tendency continues, there is a danger that the identity of environmental sociology will be diluted and it will become rootless. In other words, environmental sociology might come to consist of a miscellaneous mixture of researchers in the social sciences and the humanities with a wide range of approaches and problem consciousnesses. If its identity cannot be defined in a positive way, environmental sociology may come to be defined in negative terms— as a field of research with no particular approach. Caught between the environmental sciences and mainstream sociology, the danger that environmental sociology will suffer an identity crisis is significant.

The concept 'environment' is highly comprehensive, a rubbery concept that can be expanded or contracted. In Japan in recent years, when new faculties or graduate schools that integrate the natural and social sciences have been created in universities, the term 'environment' has frequently been used in, for example, the Faculty of Environment and Society Studies, or the Graduate School of Environmental Studies, or the Graduate School of Human and Environmental Research. There have also been cases where academics in environmental sociology and other sociological fields have attempted to bridge the gap between the natural and social sciences. The issue here is to what extent this objective has been achieved.

Identity: Object, methodology, and value interest

Where can we turn to construct a distinct identity for environmental sociology? Generally speaking, an academic discipline is determined by its (1) object or research area, (2) methodology and (3) value interests. To the degree these three elements are unique to a discipline, it has a clear identity. That 'value interests' decide the character of an academic discipline may require explanation for readers who embrace the idea of 'neutrality' in the social sciences. The value interests that define, for example, women's studies, gender studies, and minority studies are, respectively, 'the liberation of women from oppression', 'the liberation of women and men from existing gender roles', and 'the liberation of minorities from oppression'. Similarly, an important element in the identity of Marxist economics was its basic interest in the 'liberation of mankind from economic oppression'.

As an essentially reflexive discipline, sociologists tend to have greater suspicions about their discipline's identity than do, for example, economists, legal scholars or psychologists. The object of sociology is very broad, and the discipline is characterized by many competing methodologies and theoretical perspectives, such as positivism and idealism. In contrast, in economics, the 'market' is the unambiguous object, while the study of law is clearly focused on 'rights' and 'legal justice'. Understandably, some sociologists feel insecure because there is no such clear academic foundation for sociology's identity.

The objectives of environmental sociology, however, are more clearly defined. There is general agreement that the research object of environmental sociology includes: (a) environmental issues, (b) the relationship between the environment and society, and (c) society's views about the environment (or 'environmental consciousness'—the role/place of the environment in human culture). Although a strict definition of 'environment' is elusive,[16] there is some consensus that it must be centered on the natural environment and 'semi-environment'[17] and includes the historical environment and cultural environment.

In relative terms, the value interest of environmental sociology is clear: to contribute to resolving environmental issues and promoting ecological concerns. Dunlap et al.'s idea of a 'new ecological paradigm' through which environmental sociology might inform and reform mainstream sociology might be seen as a defining value interest for environmental sociology. In chapter 2 the author has proposed that environmental problems be viewed as downstream problems, and that the sociology of environmental issues should be a 'sociology of the downstream perspective'.

What then of methodology? As I argue in chapter 6, this consists of (a) an action theory perspective or a Weberian social action perspective, (b) an emphasis on the field and on field surveys, (c) an emphasis on the perspectives of residents of affected areas and other affected individuals and (d) a strong orientation towards a holistic elucidation of a problem and all of its connections. But these characteristics are not limited to environmental sociology; they are more or less common to all sociological methodologies and research. What is unique about the sociological approach is its focus on the social and historical conditions that determine values and norms, and 'structure', communities, cultures, individuals, and families as well as residents' organizations, civil groups, NGOs, NPOs, and social movements. The unique tools of sociology are field surveys designed to discover the perspectives and local understandings of victims, residents and the general public, and to identify and describe all of the complexities of an issue. While environmental legal studies and economics are technical disciplines that provide practical tools, the sociological imagination is reflexive.

Characteristics of Japanese environmental sociology

To what extent has environmental sociology been able to meet the great expectations placed upon it? To more fully appreciate its second stage of development as a discipline, it is necessary to evaluate its characteristics and achievements in concrete terms.[18]

To date the major research interests of environmental sociology in Japan have included: (1) pollution related problems, local development problems, the processes by which environmental issues emerge, and the perpetrator-victim relationship, (2) environmental movements, citizen activities, and the activities of NGOs and NPOs, (e.g., antipollution movements, and environmental conservation movements), (3) organic agriculture and social relations between organic agriculturalists and urban consumers drawing on rural sociology, (4) environmental culture, or daily life and culture's intersections with the environment, ecology, environmental perspectives and consciousness, (5) the preservation of historical houses, buildings, streets and other cultural artifacts, (6) the behavior of individuals and firms as it relates to environmental issues, such as recycling behavior, the production of waste, or efficient energy use, (7) academic theories and methodologies (such as quantitative and theoretical sociology and the theory of social dilemmas), and (8) Asia's environmental problems (especially as concerns the role played by the Japanese government and businesses). Obviously, most research projects venture into more than one of these topic areas.

To the best of the author's knowledge, (3) organic agriculture research and (5) research into the preservation of cultural heritage are not pursued by environmental sociologists overseas, but are unique to Japan. As noted above, perspectives on the social structures of victims (Iijima 1984/1993), the theory of benefit versus victimized zone (Funabashi et al. 1985), and life-environmentalism (Torigoe and Kada eds, 1984) are peculiar and excellent contributions of Japan's environmental sociologists.

Environment, Energy and Society, edited by Humphrey et al. (1982/1991) was the first representative textbook on environmental sociology published in the USA. Of nine chapters, five are devoted to global and macro resource topics, such as issues of population, energy, and food. Bell's *An Invitation to Environmental Sociology* (1998) mainly focuses on the ideologies and values that dominate nature. Hannigan's *Environmental Sociology: A Social Constructionist Approach* (1995) explores issues such as acid rain, the danger of biodiversity, and biotechnology through concrete case studies.

Comparing these texts to the earliest Japanese text in the field by Iijima (ed. 1993) or the five volume *Environmental sociology in Japan* highlights the particular characteristics of Japanese research. Japanese environmental sociology has been built on solid achievements in researching pollution related social problems, environmental destruction, citizen participation, and local communities using field surveys in local areas. It has not, however, contributed much to researching problems on a global scale, resources, or developing macro perspectives.

Some research into the problems of over-population has been conducted by Wakabayashi Keiko (2001), but there is a notable absence of research about the problems of food shortages. Very little sociological research has been conducted on acid rain or threats to biodiversity. Whaling has been researched from an environmental sociological perspective, but the only such work on biotechnology is by Ōtsuka Yoshiki (1998).

Contributions by Japanese sociologists and environmental sociologists to problems in the global environment, especially global warming, have so far been limited to works such as those by Hasegawa (1997, 1999c) and Ikeda Kanji (2001). However, the proposal to examine global environmental issues from the perspective of 'regional environmentalism' advanced by Iijima (2001: 24–6) and Terada (2001: 251–3) may prove to be a valuable contribution that only environmental sociology could make.

An important issue for environmental research in the social sciences is how to design and create a concrete road map for a sustainable society or an environmental symbiosis social system. With the understanding that environmental sociology has moved beyond its first stage of organization and systemization, and has entered a second stage in which the emphasis is on policy research, the remainder of this work will focus on the environmental movement as the creator of a new public sphere and the social mediator between environmental issues and environmental policies, analyze recent trends and potential developments, and open new perspectives for the second stage of environmental sociology.

2 Sociology of environmental issues: A look at the downstream side

Environmental issues as downstream problems

The structure of environmental problems

As many have pointed out, environmental problems are becoming increasingly diverse and widespread. Indeed, at the beginning of the 1970s, the Japanese Environmental Agency's list of environmental hazards included only seven types—soil pollution, air pollution, water pollution, noise pollution, vibration pollution, land subsidence, and smell pollution. In contrast, today, the number of pollution sources and pollutants as well as the number of types of environmental problems has increased remarkably to include for example, dioxin pollution, hormone disrupting chemicals, the destruction of natural coastlines, the failure to preserve historical places, the many species of flora and fauna on the verge of extinction, the exhaustion of natural resources and population explosions. Thus, environmental problems are becoming increasingly vast and are often difficult to perceive with the human senses alone. Their impact will continue to be felt by many generations for centuries to come, especially in cases of pollution by environmental hormones, global warming, and radiation.

Environmental problems have moved from 'industrial pollution', which became a serious political issue in Japan in the 1960s—exemplified by the outbreak of Minamata mercury poisoning disease and the three other major 'industrial diseases' that were caused by the pollutants generated in the production processes of the heavy, chemical, and mining industries and released into the atmosphere and water, causing serious physical damage to ordinary people—to the 'high-speed transportation pollution' of the 1970s, an effect of the development of the bullet trains and the construction and increased use of highways, railways, and airports. The latter included new levels of noise and vibration pollution in the vicinity of these transport routes and bases. In the 1980s 'everyday life pollution' came to our attention, referring to the negative effects that our daily activities may have (waste disposal and energy/resource consumption) on the environment. Finally, in the 1990s, acid rain, the destruction of the ozone layer by chlorofluorocarbon emissions, and other 'global environmental issues' became the focus of concern. We might thus classify pollution and other environmental problems according to the time periods in which they came under the spotlight.[1] It should go without saying, however, that in Japan today—as in the rest of the world—neither industrial nor traffic pollution has disappeared. In fact, chemical substances continue to contaminate ground waters and the soil in a seemingly endless process (Yoshida 1989, 2002). There has also been little progress in efforts to develop either technological or policy solutions to road and airport noise pollution, while the air pollution from millions of cars and trucks remains a very daunting problem. Furthermore, issues concerning the disposal of industrial and radioactive waste seem to become more and more grievous every year.

Teranishi (1997: 98), focusing on the diversity and spread of environmental hazards from geographical, qualitative, spatial, and time dimensions, proposes a classificatory scheme that divides environmental problems into three categories—'contaminating pollution', 'nature-related pollution', and 'amenities-related pollution'. Using the examples above, then, industrial pollution and traffic pollution would be categorized as 'contaminating pollution', threats to various species of flora and fauna and natural coastline destruction are 'nature-related pollution', while many types of everyday activities pollution and the failure to preserve historical housing and streets exemplify 'amenities-related pollution'.

However, over-emphasizing the diversification and spread of environmental problems might lead to ignoring the common structural factors that lie at the root of all environmental problems. It is the downstream side of environmental hazards that I aim to emphasize in this chapter.

Upstream and downstream

The terms 'upstream' and 'downstream' have come to be deployed in various fields in the sense that they are used to describe water flows. Consider the use of these terms in reference to the circulation of fuel in generating nuclear power. 'Upstream' are processes such as extracting uranium from ore, and processing it into fuel assemblies. 'Downstream' is the reuse and disposal of spent uranium fuel (Takagi 1991). We can also distinguish between the upstream side of water supply/intake and the downstream side of waste water disposal. We must note, however, that the oil industry employs these terms differently, referring to the activities prior to the extraction of oil as 'upstream', and those related to the refining and distribution processes as 'downstream'. Using the nomenclature of water and nuclear power generation processes as a guide, I propose to define the term 'upstream' to refer to the processes preceding the consumption of valuable resources—'environmental goods'—and 'downstream' to indicate the processes that follow the use of those resources, including the release and disposal of waste and other environmental burdens—'environmental bads'. Similarly, Yoshimura (1984) pioneered the use of these terms to discuss all aspects of life and argued that pollution should be viewed as a 'downstream' problem. From this perspective he refers to waste incinerators and sewage treatment plants as 'downstream facilities'.

The term of goods in economics means valuable resources, substances and services to be used for production or consumption. They are demanded, therefore treated as commodities with some positive price. 'Bads' is a new concept recently developed in environmental economics in contrast with goods (see Hosoda 1999). Typical bads are waste. Nobody demands them; therefore they must be dealt with at some negative price. If you want to pass 'bads' onto somebody else, you must pay some price, not receive some payment in exchange. So, for a long time bads were neglected in both the business/government world and the academic world. Bads are one of the key terms for thinking about a transformation towards a sustainable society. The words for goods and bads perhaps evoke things. But conceptually, goods includes services. So I propose that we consider environmental bads to include energy flows and conversions that result in environmental burdens, such as noise pollution and vibration pollution. General waste, industrial waste, radioactive waste, dioxins, environmental hormones and the emissions of gases that cause global warming constitute environmental bads. Likewise, development projects and activities that harm historical landscapes, historical places and natural coastlines should be classified as environmental bads. Thus, the definition of environmental bads used here is: 'all substances, energy flows, and activities that harm the environment'.

In this chapter, I consider the production process in its widest sense, including traffic and distribution. Doing so provides a unified perspective through which to discuss the whole of environmental problems, including industrial, high-speed transportation, and everyday life pollution and global environmental issues, as well as the 'contaminating', 'nature-related' and 'amenities-related' pollutions that I touched upon earlier. I therefore propose a new definition of environmental problems: they arise as a result of the upstream side, both production processes and everyday activities and result in the generation of environmental bads on the downstream side that must somehow be released and disposed of in the environment.

The downstream perspective and its significance

What is the significance of revising our perspective on environmental problems and treating these as downstream problems? First, as mentioned, this perspective allows us to develop a unified scheme of increasingly diverse and dispersed environmental hazards. To repeat, by emphasizing only the diversity of environmental problems we risk overlooking some of the inherent characteristics that are common to all of them.

Second, by examining the interrelationships between the upstream and downstream sides, we can analyze environmental problems produced by upstream activities. Each of us, albeit in varying degrees, contributes to the total accumulation of most types of everyday activities pollution and global warming; and we all suffer from the accompanying problems. The perpetrator-victim relationship raises questions of responsibility as well as legal, social, and moral liability. Based upon their analysis of the role of bullet train passengers in the generation of environmental bads, noise pollution and vibration pollution, Funabashi et al. (1985) developed a theory of what they refer to as the 'benefit versus victimized zone'. They also propose a classification of 'conflict of separate types' where benefit and victimized zones separate in the area and 'conflict of overlapped types' where the two zones overlap. Generally speaking, in cases of large-scale projects such as constructing and operating bullet trains, residents in close proximity to the railway suffering from huge noise and vibration pollution in victimized zones are basically separate from the passengers and other beneficiaries. It is therefore difficult to find a way to mediate and reach a consensus (see chapter 7). This contrasts sharply with earlier types of conflicts, which usually involved smaller-scale projects, where it is relatively easier to reach a consensus, because both sides can more readily understand the others' situation and position. Disputes over the construction of light-rail transit systems are typical of the conflict of overlapped types, where some people are requested to move and oppose the plan. This classification and their hypothesis that the 'benefit zone' is expanding, while the 'victimized zone' is becoming locally grievous (Kajita 1979/1988: 16) has been significant for understanding, for example, the specifics of expressway traffic pollution as a social problem, the interrelationship between the social costs and benefits when evaluating both large and small-scale projects and the major social characteristics of consensus building processes. This perspective assumes that there is a direct relationship between costs and benefits. The benefit zone comprises the organized benefits and beneficiaries (primarily on the upstream side), while the victimized zone includes both the conspicuous and inconspicuous aspects of all organized costs on the downstream side as well as all of the bearers and victims of these costs. However, analyses from this perspective may fail to explain causal relationships. It does not necessarily follow that every given benefit zone incurs an environmental cost or produces special types of victimized zone and victims. For example, generally speaking development of information technology could bring a variety of benefit zones without any serious victimized zone. In contrast, the upstream and downstream perspective enables us to identify causes and their effects, thereby revealing downstream problems as the direct results of upstream activities. This is the distinctive feature of this approach.

Third, the upstream-downstream approach is significant for making it possible to identify downstream problems at the focal point of all contemporary social issues. Before the 1960s, the main focus of social research and discourse was on the upstream side of production processes and daily activities. Later, however, with the growth and growing awareness of environmental pollution and environmental problems, it has become necessary and unavoidable to shift attention to the downstream side of society. Limitations of growth and global warming are examples of how downstream problems may decrease the scale of the upstream processes and where a radical shift in approach is most pressing. Our societies and each member needs to fundamentally change the dominant view of the environment and society that tends to be biased towards upstream problems and develop an alternative perspective that focuses on downstream issues.

In fact, intense investment in research in upstream fields, in a market that has become highly competitive, has resulted in significant technological innovations. Research in downstream fields has, however, been minimal, as investment has been restrained.

Developing new downstream technologies has therefore been difficult. For example, over the past ten years, there has been significant technological progress in the telecommunications industry, but not in technologies for curbing the noise pollution produced by motor vehicles and aircraft, or safely disposing of industrial and radioactive waste. New developments in upstream-related fields are easy to market. But solutions for downstream problems are less attractive and therefore more difficult to finance, as their benefits typically do not have direct economic value.

Risk society and 'peripheries'

Fourth, we can interpret Ulrich Beck's conception of a 'risk society' from a downstream perspective. Beck's *Risk Society: Towards a New Modernity* (1986, 1992) was published soon after the nuclear reactor disaster at Chernobyl in 1986, and became one of the best selling social science books in Europe at the time, fifty thousand copies of the German edition were sold by 1991, despite its highly abstract and esoteric style. Contemporary life is filled with a multitude of multidimensional risks, but environmental risks are the most acute and essential. Radioactive contamination from the Chernobyl reactor crossed national borders, affecting the entire world. Risk society theory assumes a critical perspective on modern times, defining contemporary society as one of increasing risks whose effects are complex, difficult to control, and often invisible to the naked eye. The argument is that we are witnessing a transformation from an industrial society characterized by inter-class struggles for the allocation of resources to a risk society in which the central struggle is about the distribution of risk. Using the terminology introduced in this paper, we can restate this argument by saying that the transformation from an industrial society to a risk society demands a change in focus— from upstream- to downstream-oriented. Environmental risks in this framework can therefore be understood as affected by the environmental bads defined earlier.

Fifth, the downstream perspective allows us to address issues of environmental justice and social surroundings, as well as social discrimination and social disparity. If we consider the downstream side from spatial and sociological perspectives, as Robert Bullard (1994) has done in his work on the conception of environmental justice and equality (cf. Toda 1994), areas populated by ethnic minorities and the very poor as well as remote areas suffering from depopulation tend to be located in the peripheries and downstream areas of a society. On a global scale, such communities tend to be concentrated in the special peripheral area known as the Third World, or more recently, the 'South'.

The terms upstream and downstream cannot always be appropriately affixed to a certain social strata or geographical location. But it is nevertheless clear that the dichotomies center-periphery and upstream-downstream often correspond to one another. Residents of the center have relatively high incomes and are better endowed with access to information and human networks. In other words, they have more resources to mobilize against environmental problems and are more independent, which makes it easier to attract alternative, environmentally-friendly industries and contribute to regional development. Moreover, it is easier for core regions to minimize their involvement in downstream activities and processes, and thus avoid related problems. The periphery, on the other hand, tends towards much higher rates of involvement in downstream activities, and already polluted regions attract new polluting sources. In some cases, the presence of one type of environmental burden contributes to the development of another, thus resulting in a double, triple, or multilayered structure of pollution.

The first rejection by legally popular vote of a proposed nuclear plant in Japan occurred in 1996 in the town of Maki (population approximately 30,000), Niigata Prefecture. The population of Maki had increased, unlike many Japanese towns in its region in the late twentieth century. This is largely attributable to its close proximity to Niigata, a large prefectural capital city (population over 500,000), to which it was linked by a major expressway and a bullet-train line, thus enabling it to serve as a commuter town.[2] In contrast, the village of Rokkasho (population approximately 10,000), in Aomori Prefecture, is situated at the northern end of Honshu and often suffers from winter blizzards and the yamase—a cold wind that blows in early to mid summer. Largely due to its isolation and harsh weather conditions, during the post-war period in which most of Japan has enjoyed tremendous economic growth, Rokkasho has repeatedly failed in its efforts to attract development projects. In short, Rokkasho has little to offer from either a historical, geographical, or climatic perspective that might be attractive to either individuals or industry. Except, apparently, for the nuclear fuel processing industry. A proposal to build Japan's first commercial nuclear fuel processing facilities in the area was accepted in 1995 as a last resort after the failure of the project to build Mutsu-Ogawara Industrial Park.[3] It has since become the location for many nuclear facilities and the site with the greatest concentration of radioactive waste in the world. In their research into the social impact of the Tōhoku-Jōetsu bullet train line construction project, Funabashi (1988: 180) notes that often 'a policy aimed at solving one problem gives birth to another problem, and the solution of that problem results in the creation of yet another one'. He refers to this process as a 'vicious chain of structural tension'. From this perspective, an examination of the social surroundings of Rokkasho reveals how the response to one risk factor can result in the creation of new risks, and that subsequent attempts to address the new risks produces yet further risks. We can schematically represent the unfolding of this process in Rokkasho as:

> failure of building settlements and cultivating farmland → the failure of the proposed Mutsu Steelworks project → failure cultivating beet roots → setbacks in developing new rice fields → the Mutsu-Ogawara Industrial Park fiasco → the introduction of nuclear fuel processing facilities → the accumulation of radioactive waste → the further concentration of nuclear facilities in the area. (see Kamata 1991)

At the time of writing, there are plans to establish more nuclear facilities, such as a MOX fuel processing plant and a second fuel reprocessing plant on the 1,500 vacant hectares of the Mutsu-Ogawara Industrial Park. We can also observe the workings of a vicious chain in typical cases of establishing nuclear power stations:

> depopulation of the area → electric utility company proposes to establish a nuclear power station → community confrontations over the issue → start of operations at the new nuclear power station → sharp drop in labor demand with completion of construction works → population decrease → local government develops a dependency on revenue from the nuclear power station → local government invites an additional nuclear power plant → construction works and start of operations → increase in the amount of radioactive waste generated → storage of this waste in the area → construction of storage facilities to accommodate spent fuel.

The cases of Rokkasho and nuclear power plants clearly illustrate how the vicious circle of compound environmental degradation operates: the establishment of a single nuclear facility or plant has resulted in

an ever-expanding number of downstream environmental problems
that have affected the whole region.

Reflecting on modern societies

Sixth, the distinction between upstream and downstream processes
can help to develop a new perspective on the modern societies that
have brought about this polarization. In pre-modern, pre-urbanized
society, the upstream and downstream were not polarized. The two
sides were unified in the realm of everyday life, constituting a whole.
Based on field research on traditional lifestyles in hamlets along the
shores of Lake Biwa, Furukawa notes that:

> before the modern water systems were introduced, water supply
> and disposal were not separate processes. It was therefore possible
> to maintain clean water. It can be said that the two could be viewed
> only as interrelated and complementary. That is because in traditional
> water management systems, if water supplies and waste were not
> scrupulously separated, to a much greater extent than nowadays, there
> was a risk that waste water would be mixed with the supply water.
> Furthermore, one person's waste water would often become somebody
> else's supply water (i.e. those who lived downstream). Thus there
> was a continuous cycle of water use that, importantly, could be easily
> observed with the naked eye. That is, the people could see that their
> waste water affected others. (Furukawa 1984: 242-3).

We can say this is a typical example of an 'environmentally
coexistent' water management system. According to Furukawa and
others, contrary to expectations, the development of modern water
supply systems removed the circulation of water from the public-eye,
thus permitting the rapid spread of water pollution (Furukawa 1984;
Torigoe and Kada eds, 1984). As contemporary water supply systems
became indispensable to urban life-styles, water circulation was
divided into separate processes—upstream and downstream. This is
reflected in Kurasawa's definition of the modern urbanized lifestyle:
'it is a communal lifestyle in which common tasks are performed by
specialized and professionalized systems' (cited in Morioka 1993:
1094; cf. Kurasawa 1977). However, mainstream academics in the
field of urban sociology, such as Kurasawa, assume that our reliance
on specialized systems is self-evident, and have stopped short of
analyzing the problematic features of this reliance or the seriousness

of its downstream effects. Such understandings of urban life tend to occlude any possibility of developing an ecological and symbiotic lifestyle that prioritizes the circulation and reuse of resources through recognition of their mutual interdependence.

In Japan as in the USA, the growing interest in environmental sociology and environmental problems was fueled by the development and spread of a new set of ecological values that revolved around a desire for a harmonious coexistence with nature, which returns the circulation of resources and their essential place in the life cycle to the spotlight. These ecological values challenged the upstream centered focus that is characteristic of the individualistic contemporary industrial society and its exploitative approach to nature, exemplified by the lifestyle common to mass-production/mass-consumption society (henceforth 'consumer society' for short). Thus, we are confronted with the difficult task of determining how to reunify the upstream and downstream processes in order to create a new system of reuse and recycling.

Seventh, conceptualizing modernity in the terms of upstream and downstream processes sheds new light on the significance and identity of environmental sociology—enabling us to see it as a discipline that focuses on the downstream side as explained below.

Environmental sociology as sociology of the downstream side

A look at the downstream side

Environmental sociology is a discipline that uses sociological methods to focus on the interrelationships between individuals, society and, primarily, the natural environment. If we apply the upstream-downstream approach, we can say that whereas mainstream sociology has primarily focused on upstream issues, such as production processes and social activities, environmental sociology is the sociology of downstream perspective and issues, such as environmental hazards, environmental culture, and environmental coexistence.

Affected by the perspective of the radical sociology movement of the 1970s, as mentioned in chapter one, Catton and Dunlap (1978, 1980) criticized mainstream sociological approaches for assuming a world-view that they identified as the 'Human Exceptionalism Paradigm' (HEP), which exclusively focuses on human societies while emphasizing social and cultural environments as the loci of human/social developments. They argued that environmental sociology should instead adopt a 'New Ecological Paradigm' (NEP) that recognizes human beings as one of myriad creatures whose very existence is dependent upon the natural environment, of which they are part, and who therefore cannot afford to ignore ecological principles.[4] That is, rather than seeing environmental sociology as a branch or subdivision of sociology—which they perceived to be a fundamentally misdirected discipline—it should be understood as an alternative to the established field of sociology. However, as also noted in chapter one, Buttel argues that the arguments supporting the 'New Ecological Paradigm' have no empirical bases and thus the concept has little value beyond its role as a rallying slogan (Buttel 1987).

I argue that the significance of environmental sociology lies in its focus on downstream issues and its conceptualization of society in terms of sustainability and resource circulation. The concerns of environmental sociology can be divided into two groups—'environmental problems' and 'environmental coexistence'. The sociology of environmental problems focuses on pollution, environmental contamination, and other environmental burdens that affect various social fields, as mentioned. The sociology of environmental coexistence examines people's environmental consciousness, environmental ethics, modes of interaction between humans and their environment, how cultural patterns of life affect the environment, and the preservation of historical places. As I have repeatedly stressed, environmental problems are those caused by environmental burdens downstream. Environmental sociology is also concerned with theoretical questions about environmental bads, the downstream side and tasks related to environmental coexistence—such as those which led Torigoe, Kada, and Furukawa to formulate the conception of 'life-environmentalism'–as well as more empirical questions such as those arising in the context of environmental preservation, or how resources actually circulate.

Social meanings of the problems of waste

In the 1960s there were four major cases of industrial pollution in Japan, each essentially an instance of painful discovery of the effects of continually releasing pollutants into the environment. In each case it was eventually revealed that the perpetrators had intentionally concealed the environmental risks of their activities. The strong public responses to these four cases resulted in action by the central government and local authorities to strengthen environmental regulations. Industry's compliance with these new regulations assisted in achieving some degree of control over environmental contamination, thus paving the way for the economic prosperity of the 1970s. As a result, a production and consumption society emerged, based on the firm assumption that safe and proper management and disposal of waste is imperative.

However, as Hasegawa (2000d) points out, the disposal of industrial and radioactive waste remains a central problem to this day, with efforts to implement environmental risk management processes that might ensure a safe society and life confronted by great technological, economic, and social difficulties. Thus while on the surface, environmental risk management appears to be well-refined, the areas where it is most urgently needed—downstream—are also those faced with the greatest difficulties, such as poor locations, insufficient or inadequate waste processing facilities, high risk of accidents and other hazards, and the sharpest splits between pro- and anti-development forces. Furthermore, in both national and global peripheries, problems such as illegal waste dumping and human errors in nuclear facilities are mounting. Location decisions are often determined by the need to strike a balance between cost and profit considerations, the necessity to maintain relationships with certain groups, and the illusory drive to develop.

Consumer society entails massive discharges of radioactive and industrial waste. However, this social form has begun to tremble as its accumulated waste has begun to eat away at the long ignored ecological underpinnings of human life.

Finding ways to convert industrial waste into usable resources while attempting to minimize the amount of waste generated and discharged—that is, developing a sustainable society—is our goal for the future. Only through such efforts can we hope to eliminate the global problems of climate change and progress toward a world that no longer deploys nuclear power, and thus no longer generates the radioactive waste that will undoubtably affect countless future generations.

From activism inspired by environmental damage to policies inspired by activism

The sociology of environmental problems is based on a four-fold distinction: perpetrators, victims, movements, and policies. Many academic disciplines that address environmental problems and environmental policies—especially environmental economics and environmental law—strongly emphasize policies. Environmental sociology, in contrast, has developed a holistic perspective on downstream problems, particularly as they affect particular members and groups of the community in their everyday lives. The perpetrators' perspective attributes responsibility for these problems to the actions and actors upstream, while the downstream perspective reveals the victims who suffer from pollution, allowing fuller analysis of the social costs and hardships of pollution.

Environmental activism—social movements motivated by environmental problems and the promotion of environmental safety—at its best serves to mediate between perpetrators, victims, and policy-makers. It can also be seen as a protest voiced from downstream. 'Without the environmental movement of the late 1960s, environmental sociology would probably not have emerged' (Humphrey and Buttel 1982: 7). In both the USA and Japan, the fact that the academics who specialize in the study of environmental activism have been the ones to secure leading positions in the field is indicative of the strategic importance of environmental activism. By addressing questions such as 'How will public works projects affect the life of the community in question, and what kind of environmental problems will they lead to?' and 'In what ways and based on what logical arguments can local residents' campaigns and citizen's movements as well as environmental NPOs demonstrate their protest?' from the perspectives of those who have suffered from the effects of pollution, environmental activists and the academics who study them have defined the central battle ground for environmental sociology.

At present, places such as Kawasaki, Minamata, Nishi-yodogawa Ward in Osaka City, and Mizushima District in Kurashiki City, places that have suffered from the negative effects of pollution, have seen the development of a new type of activism—one that proposes some solution rather than 'merely' protesting against pollution. Such movements have been formed on the initiative of local residents and local governments and are aimed at restoring the environment.[5] In addition to strengthening anti-pollution measures, these movements promote environmental education, encourage recycling, tree-planting and eco-business activities, and attempt to move their local area away from the downstream side of consumer society.

It may seem that in terms of affecting policy, environmental sociology lags behind environmental economics and law, but the achievements of environmental activism indicate the opposite. The areas in which activists' and policy-makers' perspectives overlap offer the greatest potential for developing new ideas unique to environmental sociology. Environmental sociology should therefore move from activism in reaction to environmental damage to a policy-making orientation inspired by activism. But as discussed in ensuing chapters, guided by the practical orientation of Japan's environmental activism and its focus on everyday life, environmental sociology in Japan has the potential to have a unique impact on environmental policy.

Sustainable development is one of the prerequisites for achieving a sustainable lifestyle and society. How can we integrate the upstream and downstream sides, so that they are once again understood to be a unified system? Developing concrete policies based on analysis of the downstream side will require sociological imagination. This is the tremendous task before us.

Part II:
The sociology of environmental movements

3 Environmental issues and movements

Introduction

Environmental movements are strongly influenced and shaped by the structures of the particular environmental issues they face and the associated perpetrator-victim relationships. In this chapter I first examine and compare the different characteristics of residents' and citizens' environmental movements in the Japanese context. We will see that the Japanese environmental movements have developed primarily as residents' movements. After comparing the structures of industrial pollution, high-speed transportation pollution, everyday life pollution, and global environmental issues, I will return to the structure of each to reveal how their historical changes have affected the environmental movements. Of the looming global environmental problems, the most prominent is global warming. The affects of global warming are largely invisible, and it has special characteristics that complicate the development of a residents' or citizens' movement. Movements concerned with global warming are therefore primarily driven by highly specialized and professionalized environmental NGOs acting on an international stage.

Residents' movements and citizens' movements

The everyday nature of residents' movements and citizens' movements today

Local communities are facing a number of environmental issues, including the construction of golf courses and resorts, the protection of forests, garbage disposal, waste disposal, recycling, nuclear power reactors, the preservation of historical streets and houses, the consolidation of parks, pollution by agricultural chemicals, noise pollution from airplanes, cars and trains, air pollution from tobacco smokers, the right to sunlight, and others. It would be difficult to not be aware of the fact that each of these issues is being tackled by some citizens' or residents' movement, or that such groups are engaged in a range of activities. Local newspapers and the local pages of national newspapers often contain announcements of such activities and meetings. Local Community Centers and health food shops frequently display such announcements or display free information leaflets and fliers. Today, rallying residents' or citizens' movements to tackle environmental issues has become an ordinary part of everyday life in many local communities in Japan. For instance, a 1991 guidebook of citizens' movements and networks in Miyagi Prefecture included contact details for 400 citizens' groups including 75 identified as environmental groups,[1] a rough indication of the extent to which organized environmental movements have become part of everyday life. More recently, it has become increasingly common to see groups organize themselves as environmental NPOs.

What are social movements?

As will be discussed in more detail in chapter 5, from the perspective of social movement theories, social movements are generally defined as 'collective actions oriented towards change that are based on grievances and discontent with the status quo or an anticipated state of affairs'. The key terms here are 'discontent', 'orientation towards change', and 'collective action'. An essential characteristic of a social movement is discontent with or a sense of deprivation arising from a particular issue. And it is not sufficient that this sense of deprivation or dissatisfaction be experienced by one or two individuals; it must be a collective experience commonly shared by some group of people in similar circumstances. There are many social movements concerned with a wide variety of issues, such as the peace movement and the women's movement, but my focus here is limited to social movements in Japanese society that are organized in response to environmental problems.

Analysts differ sharply in their views about the correlation between levels of discontent about an issue and the emergence of a social movement to address it. Some consider discontent to be an important variable for explaining the rise of a social movement (e.g. the theory of relative deprivation; see Geschwender 1968; Gurr 1970) while others argue that it has no validity as an explanatory variable (e.g. the early version of resource mobilization theory; see McCarthy and Zald 1977). In fact, its explanatory value is itself a variable, dependent upon context. My hypotheses are as follows: When the deprivation has been relatively stable historically, the ability of discontent to explain the birth of a movement is relatively small. But when the deprivation arises from relatively recent and unstable circumstances, especially where the issue is such that a rapid increase in the level of deprivation is foreseeable, discontent may prove to be a central factor in explaining the birth of a new social movement. In more concrete terms, for movements demanding the extension of rights to groups and individuals who have never before enjoyed them, such as the civil rights movement, dissatisfaction is a relatively weak causal factor—their historical duration demands the question 'why now?' to which an answer can only be found among other variables. But in social movements that have risen in response to environmental problems, where people are not demanding new rights but rather the protection of established rights that face new threats, dissatisfaction is a much stronger explanatory factor.

Offe identifies four key features to be considered when discussing social movements: (1) the actors, (2) the issues, (3) the value-orientation of the movement, and (4) the mode of action (Offe 1985; Hasegawa 1990). When we refer to the peace movement or the women's movement, the civil rights movement or environmental movements, we are identifying the movement by its value-orientation. If we want to understand how social movements work, though, it is useful to identify different types of movements by their actors. From this perspective Japanese environmental movements can be roughly categorized as either residents' movements or citizens' movements. Residents' and citizens' movements are defined by their actors, who are either 'residents' or 'citizens', and protest from their positions as such. Both types of movement have played important roles in promoting the social recognition of, social interest in, and shaping public opinion about environmental problems, as well as driving policy changes and the development of various means to tackle these problems.

Their characteristic mode of action is distinct from the broad focus of, say, a political party or labor union, a neighborhood association or municipality (although these types of organizations sometimes form the nucleus for, and often provide support to, residents' movements). They generally take the form of residents' organizations, citizens' organizations, or looser networks that are oriented towards resolving a specific issue. Many, in recent years, have been organized as NPOs.

'Residents' as interested parties and 'citizens' as 'conscience constituents'

What then is the difference between a residents' movement and a citizens' movement? The two terms are commonly used without distinction, almost interchangably. Indeed, the difference between them is not always entirely clear. For example, Miyamoto (1971: 2) defines a residents' movement as 'a movement where residents have a set of demands or a problem, a residents' organization to address this problem, and approach the government, municipal authorities or firms to solve the problem'. If we were to replace each appearance of the term 'residents' with 'citizens', we would also have a formal definition of citizens' movements. We must therefore examine the qualitative differences between the concepts 'residents' and 'citizens' if we are to more properly distinguish between the two types of movements.

Of course, there are subtle nuances between the terms 'resident' and 'citizen'. Residents are people with close connections to particular local communities, while in the context of social movements citizens are expected to be autonomous individuals.[2] Contrasting residents' and citizens' movements from this perspective reveals important differences in the character of the respective movements and their organizing principles. Table 3.1 presents a simple schematic comparison of some of the key differences between the two types of movement.

In general, residents' movements are commonly organized around existing local groups such as neighborhood associations, and are strongly characterized by their focus on issues of concern to a specific locale of relatively limited geographical range (such

Table 3.1 The basic character of a residents' movement and a citizens' movement

	Residents' Movement	Citizens' Movement
Actors		
a) Character	Residents as interested parties	Citizens as conscience constituents
b) Hierarchical background	General citizens, farmers and fishers, self-employed, women, the aged	Professionals, highly educated
Issue	Protect or achieve specific interests in neighborhoods	Protect or achieve universal values
Value orientation	Particularism, limiting	Universalism, autonomy
Modes of action		
a) Build up ties with	Neighbors in local community	People with communal beliefs and values
b) Focus of action	Instrumental rationality	Value orientation
c) Qualification for involvement	As neighbors	As supporters

as a local school district). Typically, their membership is largely comprised of the people who reside in this limited locale. In contrast, citizens' movements consist of autonomous individual citizens who share ideals and/or objectives. They are strongly characterized by their focus on much broader issues; issues of concern to entire cities or prefectures, and sometimes of concern to all of humanity (e.g. the peace movement and women's movement). The hierarchical background of either can vary greatly, but the primary actors in a residents' movement have a strong interest in the local community, and are often people who find it relatively easy to devote time to the movement, such as those involved in agriculture or fishing, the self-employed, public sector employees (i.e. public officials and teachers) in urban areas, women and the elderly. In contrast, the people involved in citizens' movements are typically professionals and the better educated who have access to information resources.

For instance, if a society to protect the natural environment of 'Mt. X' was organized by or for those with a direct interest, i.e. people who live in the immediate vicinity of the mountain, this society would assume the characteristics of a residents' movement and the central objective of the society would be to protect the interests of the local people. In contrast, if instead of local people, the society was created by the general citizenry of surrounding urban areas, lovers of nature, and specialists such as scientists, teachers and lawyers, it would assume the characteristics of a citizens' movement, and its activities would be much more likely to be pedagogic and idealistic. There is a strong tendency for the 'citizens' in a citizens' movement to be 'conscience constituents' who participate from a commitment to specific values rather than direct interests as McCarthy and Zald (1977) describe.

Of course, residents' movements are frequently supported by 'citizens' from outside of the local area, and often it is the residents who embrace the ideals and principles of broader citizens' movements who play the leading and/or central support roles in a residents' movement. In fact, cooperative links between residents' groups and citizens' movements are quite common in cases where a high degree of specialist knowledge about natural science or technology is required by a protest movement; for example, where movements oppose the construction of massive industrial complexes or nuclear power stations, or movements are fighting to protect the natural environment. More concretely, Japanese residents' movements that oppose nuclear power-stations and fuel processing facilities have been strongly supported by citizens' movements in nearby urban areas or large metropolitan areas. Where a residents' movement has remained active and organized for a long period of time, it is often supported by a citizens' movement. As we will discuss shortly, the particular structure or form assumed by an environmental residents' or citizens' movement is largely determined by the particular characteristics of its perpetrator-victim structure.

A history of residents' movements and environmental issues

The correspondence with postwar history

Japan's postwar history is often considered in ten-year blocks defined by economic developments. Major developments in the history of Japanese environmental movements roughly correspond to these developments. Economic histories of postwar Japan typically identify: the postwar recovery period from 1945 to 1954, the first stage of rapid economic growth from 1955 to 1964, the second stage of rapid economic growth from 1965 to 1974, a period of relatively stable growth from 1974 to 1985, and the period of post-industrialization characterized by the spread of information technology and internationalization from 1986 to the present.

I shall elaborate briefly on the turning points for each of these periods. The end of the postwar recovery period is marked by the beginning of the so-called '1955-system', referring to the year when the conservative parties in Japan came together and formed the Liberal Democratic Party. This was the beginning of a long period during which a single party dominated Japanese politics, and economic growth was the primary focus of political activity. This period came to an end in 1964, coincidentally the year that Tokyo hosted the Olympic Games. The Tōkaidō bullet train line was opened in 1964, and the word 'pollution' began to permeate public discourse in the wake of the publication of Shōji and Miyamoto's *Awful industrial pollutions* (*Osoru beki kōgai*) (1964). During the ensuing decade the problems of pollution and environmental destruction came to be recognized as the shadows and distortions of rapid economic growth and became full-blown social issues. The oil crisis of October 1973, when oil exporting countries in the Middle East restricted supply and caused an explosive rise in oil prices, brought an end to this (global) period of high economic growth. Although there is no commonly accepted date for the end of the post-oil-shock period, 1986 proves convenient as the end of another decade. More importantly for our purposes, however, in 1986 the significance of global environmental problems burst into the global public consciousness following the nuclear accident at Chernobyl. The ensuing decade saw a sharp increase in the number of social movements around the world, a development that strongly shaped the post-1990s world. This period is characterized by the dialogue between the East and West that soon led to the end of the Cold War, a sharp and rapid increase in the international value of the yen, and the beginnings of the 'bubble economy'. I will return to the global environmental issues of this era shortly.

Table 3.2 compares and summarizes the characteristics of the perpetrator-victim structure and movement of the four types of industrial pollution, high-speed transportation pollution, everyday life pollution, and global environmental issues.

Residents' movements, prehistory and the era of high-economic growth

According to Iijima (1984, 2000), farmers' began protesting about and resisting the water pollution caused by mining during the Edo period, in the process pioneering a form of residents' movement. The Ashio copper mine incident during the Meiji era provides a classic example of this early form of environmental movement. With the heavy industrialization of the post-Meiji period, the sources and types of pollution became much more diverse, but prewar residents' movements were sporadic and highly localized because the prewar Great Japan Imperial Constitution substantially limited civil liberties like the freedoms of speech and assembly, there were hardly any protest activities except for political parties' activities, labor movements and religious movements. It was not until after 1955, when policies to promote high levels of economic growth were introduced and heavy industrialization rapidly developed, and after democratization and liberalization, that residents' movements and groups could be found across the entire nation.

The ever-increasing rates of pollution of air, water and soil by industrial waste—exemplified by the 'four major' cases of industrial pollution[3]—stimulated widespread protests and social action from residents' groups across the country. Thus it is only from the late 1950s and early 1960s that we can legitimately speak of a full-fledged residents' movement in Japan. During this period residents' movements were strongly characterized by a tendency to seek relief

Table 3.2 Characteristics of the four types of environmental problems in Japan

| | Types of pollution | | |
Industrial	High-speed transport	Everyday life	Global environmental issues
Time period			
From 1960s—high economic growth	From 1970s—stable economic growth	From 1980s—stable economic growth in post industrial society	From 1990s—post Cold War
Examples and types of environmental degradation			
Four major industrial incidents, water, soil, air, noxious odors	Airport noise, military air base noise, bullet train noise and vibrations, highway noise	Detergents, spiked tires, household garbage, noise from neighbors, sun shading from high-rise buildings	Acid rain, forest degradation, international waterways, destruction of ozone layer, global warming, nuclear hazard
Origins of names			
Damage	Traffic facilities	Commodities	Extent of damage
Major damage			
Heavy metal poisoning	Traffic noise, vibration, stress-related health issues	Environmental degradation, human health	Irreversible international and intergenerational degradation
Primary loci			
Heavy industrial production processes	High speed traffic services	Everyday life of ordinary citizens	Production processes and everyday life
Direct causal factors and mechanism			
Discharge of untreated effluent and waste	"Un-buffered" proximity between transport facilities and houses	Agglomeration diseconomies	Agglomeration diseconomies, limit to environmental burdening, limit to growth
Structural background			
Pursuit of economic development, interest-seeking corporate behavior, absence of responsible regulatory office	Large-scale development projects, speed, government monopoly on definition of public interest, absence of effective and responsible regulations	Mass consumption, weak sense of public morality, absence of effective and responsible regulation	Mass consumption, invisibility of issues, myth of growth, absence of effective and responsible regulatory office
Major perpetrators			
Heavy industry	Government planners, supplier of traffic services	Ordinary citizens and general customers	General industry and consumers
Major targets for environmental movements			
Polluting companies, responsible regulators	Supplier of traffic services, responsible regulators	None	Developed countries
Major victims			
Nearby residents	Residents neighboring traffic service	Ordinary citizens	Future generation
Type of perpetrator-victim relationships			
Separate	Separate (expanding benefit zone and local grievous victim zone)	Overlapping (expanding beneficiaries and sufferers)	Separate (international and intergenerational distance)
Types of movements			
Oppositional movements by victims and supporters	Oppositional movement by victims	Campaign for self-restraining consumption	Collaboration between NGOs and government
Separate (expanding benefit zone and local grievous victim zone)			

and improvement after the fact, primarily through demands and protest actions in support of appeals and petitions from farmers and fishermen who were concerned about the destruction of their means of livelihood by pollution.

In Japan, residents' movements began in response to environmental pollution, and pollution in turn forged the character of residents' movements. Here, residents' movements differ from citizens' movements in their primary focus, the early examples of which frequently revolved around struggles for peace, such as the Ban the Bomb movements, protests against the 1960 US-Japan Security Treaty, and the anti-Vietnam War movement.

Large-scale development and the industrial pollution type residents' movement

The residents' movements in the second stage of high economic growth (from the mid 1960s) had four distinguishing characteristics. First they were typically protests against large-scale development projects, especially the construction of industrial complexes that would produce industrial pollution. A good example of this was the successful 1964 movements against the construction of petro-chemical complexes in Numazu, Mishima and Shimizu where residents were anxious about severe damage. The organizing processes, strategies and tactics of these movements had a significant impact on their successors. Second, the actors in these movements came from across Japan's social strata. The 1964 example once again provides a good illustration: in a relatively short period of time housewives joined with teachers, researchers and other professionals, plus a broad spectrum of local residents including laborers, farmers and fishermen, and stood shoulder to shoulder to defend their local environment. Third, social movements during this period typically involved trans-regional and/or inter-organizational cooperation. In the example cited this manifested as cooperative efforts between movements in three neighboring urban areas. Fourth, the general public—and therefore the residents/activists—were far better informed than their predecessors had been. Sticking with our example, at an early stage of this movement local residents were already aware of the pollution and associated environmental destruction in other places where large industrial complexes had previously been built, such as Yokkaichi, Chiba and Mizushima, and were thus well-prepared to reject and counter the developmental illusion that the construction of an industrial complex would be good for local development.

It is worth noting that these movements were not simply responses to the direct actions of individual corporations (although individual corporations were often the direct or explicit focus of protest, criticism and litigation), but rather were reactions against government policies. This period of high economic growth was marked by the introduction and (attempted) implementation of several nationwide development plans, including the First Comprehensive National Development Plan (1962) which aimed to develop and disperse industrial activity across the nation, in part by designating and funding new industrial cities, and the Second Comprehensive National Development Plan (1969) which attempted to stimulate development across the nation through the construction of enormous industrial complexes and interconnecting transport networks of bullet train lines and high-speed freeways. In the Japanese context then, the term 'large-scale development projects' sometimes refers to the type of development outlined in the Second Comprehensive National Development Plan, but in the Japanese context more generally it means: 'state-led development projects in which the government and government agencies play a major directing role and where massive amounts of public funds are invested' (Hasegawa and Funabashi 1988: 4). The government embraced these large-scale development projects as a means to maintaining the high rates of economic growth that Japan had been enjoying. Economic growth was seen, in turn, as a means of resolving the problems of over-population and depopulation, urban issues, and regional disparities.

However, the massive scale of these developments ensured that their (drastic) impact on the natural and social environments surrounding the industrial complexes was almost immediately apparent. The public awareness of the four major pollution cases, the pollution problems evident in other industrial areas, and the widely-reported success of the opposition to industrial complexes in Numazu and elsewhere stimulated the growth of further residents' movements all over Japan to oppose large-scale developments. Powerful residents' movements were seen in a number of areas—such as Tomakomai City in Hokkaidö (Motojima and Shöji eds, 1980), Rokkasho Village in Aomori (Funabashi et al. eds, 1998), and the Shibushi Bay area in Kagoshima—when it was revealed that they had been identified by the Second Comprehensive National Development Plan as sites for new large-scale petrochemical complexes. Public opposition to these plans was so fierce that in the aftermath of the first oil crisis, as governments around the world reassessed their energy supply options and associated economic policies, the Japanese government abandoned its plans for further petrochemical industrial complexes altogether. Broadbent (1998) provides a typical example with his case study of a New Industrial City project that was suspended and finally abandoned due to the protest movements.

The political effectiveness of the residents' movement

The political effectiveness of exercising the citizens' right to protest received great attention from the media and general public. Protesting was embraced throughout Japan as it came to be seen as a tool for resisting government policies that valued economic growth above all else and lacked any mechanisms for effectively regulating pollution or preventing environmental destruction. From the mid 1960s, these movements broadened and became more active in demanding improvements to the living environment. Urban dwellers who had developed a citizens' conscientiousness began to organize in defense of the environment, to create new communities and to revitalize towns. In the process, the foci and forms of the residents' movements became much more diverse, and the differences to citizens' movements became far more ambiguous.

Especially in and around large metropolitan areas such as Tokyo, Osaka and Nagoya, the provision of social capital had not kept up with the population explosion of the high economic growth years. Failures in water supply, sewage, road-works, employment and educational opportunities, public health, and maintaining parks and other public facilities were highly visible. Discontent with these changing circumstances was an important factor in the nation-wide boom in progressive municipal governments and the political behavior of progressive voters in the large metropolitan areas in the late 1960s and early 1970s. Minobe Ryōkichi's election as Governor of the Metropolis of Tokyo in 1967 was a seminal event for this movement.

High-speed transportation pollution and conflicting views of the 'public'

The Second Comprehensive National Development Plan was based on two pillars: constructing massive industrial complexes, and a high-speed national transportation infrastructure. This second pillar soon produced 'high-speed transportation pollution', an environmental problem that is quite distinct in many ways from industrial pollution, and a new type of problem for local communities to come to terms with.

During the later years of high economic growth (the early 1970s), high-speed transportation pollution became a central issue for residents' movements, especially in the large metropolitan areas. Representative examples include movements to: rectify the Osaka Airport pollution problem; oppose the construction of Narita Airport; rectify the Nagoya bullet train pollution problem; oppose the construction of the Tōhoku and Jōetsu bullet train lines in Saitama Prefecture and northern Tokyo, and; oppose the construction of the Yokohama cargo line. In all of these cases, large groups of residents were successfully organized in a relatively short period of time, and an organized and powerful residents' movement was active for a long period of time. Each of these cases had a significant impact on subsequent residents' movements, transportation policy and the judicial system. The Nagoya bullet train Pollution Response Alliance organized 2,000 households residing within about 7km of the train line. The anti-bullet train movement in Urawa demanded that the line be built underground and at its peak could mobilize about 4,000 people to a mass rally (Funabashi et al. 1985; Funabashi 1988). In other cases, groups of plaintiffs were organized from residents' movements; for example, in litigation over the pollution produced by Osaka Airport and the Nagoya bullet train line, where the suspension or prohibition of airplanes and trains was sought to reduce the environmental damage. Alongside the litigation over the four major industrial pollution cases, these cases helped to establish a kind of 'class-action' lawsuit as an effective strategy for residents' and citizens' movements, and exposed the limited capacity of the judicial system to provide relief (Hasegawa 1989 and chapter 7 of this book).

Legal action by residents' movements seeking redress against industrial pollution principally indicted the interest-seeking behavior of private firms, accrediting responsibility to them. In contrast, in the airport and railway cases mentioned above, the then Ministry of Transportation and the public bodies in charge of designing and running transportation facilities—i.e. the national railroads and airports—were targeted for their refusals to accept responsibility for the harm they were causing, their refusals to implement effective policies to prevent noise pollution, and their continuing 'defense' of their destructive behavior in the name of 'the public'. Here, the public actors/perpetrators responded to the residents/victims with attitudes and tactics that were chillingly similar to the corporations' responses to accusations of industrial pollution. The government and public agencies define 'the public good' according to the social usefulness of public works and social capital, and used this to justify the behavior of public authorities, proceed with the construction of these projects, and as grounds for imposing restrictions on private rights to reduce (or avoid) downstream damage and harm resulting from their activities. Residents criticized the authoritarian and oppressive tone of this conceptualization of 'the public', arguing that the public good ought to be defined through democratic processes, including a complete prohibition on any violation of fundamental human rights, and the demand for informed-consent of residents in surrounding areas through direct participation in the decision-making processes before any large projects proceed.

'Highly advanced' mass consumer society and everyday life pollution

Bullet trains, jet aircraft and high-speed freeways are essential to highly industrialized modern urban life, greatly reducing the time and cost of moving goods and people over large distances. Noise pollution from bullet trains and jet aircraft became social problems as the number of flights and trains rapidly increased in response to growing demand. Efforts by the government, private enterprise and the mass media to shape and create demand played a significant role. As the highly technologized mass consumer society continues its advance, the accompanying environmental problems extend well beyond high-speed transportation pollution. The seemingly unending quest for convenience and greater simplicity characteristic of mass consumer society has generated a wide variety of new everyday life pollutants, including detergent pollution, household garbage pollution, noise pollution from neighbors, and spiked tire pollution. Issues involving everyday life pollution came to characterize residents' movements from the mid 1970s era of stable economic growth. The primary characteristic of everyday life pollution is that it generates severe environmental destruction through the ordinary consumption behavior of general citizens. Although the individual impact each citizen has on the environment is negligible, cumulatively they create severe environmental and social problems. Research by Umino and others on 'the social dilemma' neatly models this mechanism.[4] Here, the perpetrator can also be the victim, as in the cases of the pollution caused by spiked tires and household garbage pollution.

Spiked tires have metal studs embedded to prevent cars from slipping on icy roads. They are highly effective, and unlike chains do not require any effort to put on or take off. In the 1970s they came into wide use very rapidly, especially in snowy northern Japan. But when used on dry roads, they tear up the road surface and create dust. Thus in the late 1970s and early 1980s they became a serious environmental problem in areas such as Sendai, Sapporo. In Sendai, a broad movement to encourage voluntary restraint in the use of spiked tires developed, including cooperation between public authorities, the mass media, residents' groups, and the local lawyers' association. In Miyagi Prefecture, in 1985 restrictive ordinances were first enacted at the prefectural level. Similar ordinances were subsequently enacted by Sapporo and other prefectures. This led to nation-wide legislation in 1990 that prohibited the use of spiked tires altogether (Hasegawa 1998c). In other cases, public administration officials have taken the lead, working with residents' groups to sometimes promote, sometimes discourage the use of certain products, as can be seen in the movement to abstain from using phosphorous synthetic detergents and use soap powder around Lake Biwa which started in 1975 (Katagiri 1995; Wakita 2001).

In these cases, the possibility of resolving the issue through the use of alternative technologies or alternative products was high, and the concerned industries were not antagonistic to these movements. From a relatively early stage the industries endeavored to develop and market alternative products (e.g. studless tires, four-wheel drive vehicles, and non-phosphorous synthetic detergents). In situations where the perpetrators are also the victims and the movement is not facing a powerful adversary, it is relatively easy to reach a social consensus (although not necessarily an adequate solution—for example, the introduction of non-phosphorous detergents did not entirely resolve the problem of detergent pollution, but merely the issue of phosphorous in detergent pollution). These easy consensuses make such situations very attractive for politicians and public officials to claim the initiative. Issues surrounding garbage reduction, collection and recycling have similarly long histories, and similar structural characteristics.

Residents' movements and citizens' movements in the 1980s

From the mid 1970s, after Japanese society entered a period of stable growth, there were fewer opportunities for the mass media to focus as much attention on the residents' movement as they had in the past. First, drawing on the experiences of industrial pollution in the 1960s, from around 1970 various legal mechanisms were gradually put in place to prevent pollutants from being released into the environment and the situation slowly began to improve. One of the most significant and far-reaching changes at this time was the removal of the 'economic harmony clause' from government regulations and legislation. The 'economic harmony clause' had stipulated that 'with regard to the protection for the living environment, harmony is to be maintained with the healthy development of the economy'. During this period the clause was removed, for example, from the Basic Law on Pollution Control (1967) when it was revised in 1970. The Law for the Settlement of Environmental Pollution Disputes (1970), the Pollution-Related Health Damage Compensation Law (1973), and new environmental standards were established without it, as were municipal level agreements on standardized methods of preventing pollution.

Second, as a result of slowing economic growth and financial difficulties, the number of new large-scale development projects has dropped and many others were effectively abandoned. Third, there has been a qualitative change in the nature of the issues of concern: from industrial pollution and high-speed transportation pollution where the relationship between perpetrators and victims is clearly conflictual, to everyday life pollution where it is much more difficult to identify an adversary. Of course, in issues such as nuclear energy where those in favor and those opposed are clearly distinguishable and the opposing parties have no effective opportunity to negotiate, confrontational-type disputes still arise. In the 1980s confrontational movements actively opposed the construction of an airport on Ishigaki Island (Ukai 1992) and the construction of the US military housing complex at Ikego forest (Mori 1996).

I must stress, however, that the fact that residents' and citizens' movements are not receiving the media attention they once did does not mean that their influence has dissipated. Today there are many residents' and citizens' movements that have endured for more than ten (and sometimes twenty) years. As these movements have accumulated knowledge and experience, they have become part of everyday life. Indeed, as previously mentioned, there have been cases where administrative authorities have initiated residents' or citizens' organizations in efforts to mobilize the voluntary cooperation of residents. Social action by residents' and citizens' groups have become well established (and socially acceptable) means for residents to voice their objections to proposals affecting their local environment, and to otherwise participate in political decision-making.

In a discussion about the birth of environmental sociology in the USA, Humphrey and Buttel note that 'Without the environmental movement of the late 1960s, environmental sociology would probably not have emerged.' (Humphrey and Buttel 1982: 7). Similarly, without the residents' and citizens' movements discussed above, Japan's unique contribution to environmental sociology would not have developed. Residents' and citizens' movements have helped to define how sociologists view environmental issues, and in the process have shaped and nurtured environmental sociologists. Today such movements continue to define and inform many of the issues that environmental sociology needs to discuss. I will return to this discussion in chapters 4 and 13 while examining the transformation of citizens' movements into NPOs in the 1990s.

Global warming issues and the environmental movement

The invisibility of global warming

Beginning in the mid 1980s, and intensifying after the end of the Cold War, global environmental problems came to the fore in Japan and around the world. Global warming is perhaps the most prominent of these, and has become the focus of much attention and heated debate. In this section I will outline the distinctive characteristics of global warming and the particular difficulties facing the corresponding environmental movement.

Generally speaking, responses to environmental issues are stimulated by (1) the visibility of the issue and its victims, (2) the urgency for a response, and (3) the ease of rectifying the situation, whether through technical measures or strategic counter-action. When an issue is not highly visible, a response is not identified as urgent and/or technical and strategic responses are not readily available, it is far more difficult to develop an effective response.

The spatial and temporal dimensions of global warming are immense—it will affect the entire world, and arguably, its influence will have a duration of 50–200 years. All productive activities, all social activities, and indeed many life forms—including human beings—produce carbon dioxide. Hence, global warming might be considered to be the greatest, most far-reaching and most fundamental issue faced by human beings today.

But the problem of global warming is difficult to comprehend because the issue exceeds the 'human scale'. Its spatial and temporal scale is enormous, its multi-faceted causal connections are incredibly complex, and its potential ramifications extend into the fundamentals of life itself. Paradoxically, its overwhelming enormity tends to render global warming somewhat invisible; it exceeds the experience of everyday life and the power of the human imagination.

The first difficulty with trying to comprehend global warming is the absence of a specific 'locale'. Although some analysts predict that the small island nations near the equator will be severely affected by rising sea levels and other climatic changes, at present there is no 'local place' where acute damage is occurring. A report by the then Environmental Agency of Japan, entitled 'The serious impacts of global warming' (Chikyū ondanka no jūdai eikyö) concluded that the affects will be felt in many different ways, including water supply shortages, flood damage, an increasing number of species becoming extinct, smaller rice and corn crops, the disappearance of sand beaches, the submersion of lowlands, higher mortality rates for older people and a heavier photochemical smog. But none of these impacts will be concentrated in specific locales.

To use a concrete example, in Sendai, if the sea level rises by about one meter, the impact of the new tide is anticipated to extend to two meters above the existing sea level, which estimates suggest will damage approximately 30% of the Wakabayashi area through frequent flooding. And yet, the residents of Wakabayashi today do not appear to be experiencing any heightened sense of crisis. Nor do residents of the low-lying parts of Tokyo who will be similarly affected.

Second, for most people the future one century from now is a future beyond imagination. Reflection on the advances made in scientific technology over the past century somewhat understandably leads many people to believe that the problems facing us—such as industrial and radioactive waste disposal and global warming—will eventually be resolved by further technological developments. Needless to say, there is a psychological component in this process through which continuing current consumption patterns and lifestyles with their high energy demands are justified, or at least excused. This psychological element increases the already great difficulties in trying to convince the general public of the urgency of taking the action necessary to redress global warming.

Third, carbon dioxide cannot be seen. At the same time we 'know' that every exhalation by each animal on the planet releases carbon dioxide into the atmosphere—day in, day out, week after week etc. But this is abstract knowledge. It is not immediately experienced and thus not a conscious part of the everyday lives of ordinary people. So too, the manifold uses of fossil fuels in every aspect of everyday life, such as automobiles and electricity, to name just two obvious examples. Even when we understand intellectually that turning on the air-conditioner increases carbon dioxide emissions, it is not the type of thing that one can directly 'feel'. Extending this example, even when equipped with the information that reducing the temperature in a heated room by one degree will reduce the amount of carbon dioxide emitted by each household by 17kg per year, we have no sense of what a kilogram of carbon dioxide in the atmosphere looks or feels like, nor of its affects etc.

In the case of the dust pollution created by using spiked tires, once the cause was identified, people became much more aware of how they were personally affected by the issue. One could see at a glance which cars were and which were not equipped with spiked tires, and the distinctive sound they made announced their presence to everyone they passed. In other words, the victims could both identify *as* victims *and* identify individual perpetrators. Later, as usage declined, the improvements in air quality in central Sendai were discernible with the naked eye, and thus ordinary citizens could 'feel' that the measures taken were having an effect.

In contrast, with global warming it is difficult to identify who is and who is not cooperating. Neighbors may keep an eye on how much garbage each household is producing, for example, but it is difficult to determine who is producing how much carbon dioxide, or who has genuinely reduced their energy consumption. And individuals are frequently not aware of the affects of their own behavior.

It is not an exaggeration to say that, in relationship to future generations and the peoples of the peripheries, the populations of advanced nations today are 'free riders' (see chapter 4). Therefore, the issue of global warming is one where direct regulations such as prohibitions or restrictions on polluting activities may not necessarily be effective, but indirect regulations such as environmental taxes or carbon taxes might.

Fourth, although on current evidence the most important action required to ward off global warming is to reduce fossil fuel consumption, as mentioned, many believe this can be achieved through the development and diffusion of new technologies. However, because we cannot predict the development of effective alternative technologies (such as studless tires to replace spiked tires or fluorocarbon-free manufacturing processes), as efforts continue to develop and diffuse electric and hybrid cars, wind and solar power technologies and fuel cells etc, there appears to be no other option in the long term than to continue endeavors to reduce energy consumption.

Fifth, it is difficult to identify a particular perpetrator of global warming. Unlike the corporations responsible for industrial pollution, there is no specific organization that can be identified as *the* perpetrator, although particular industries, such as automobile manufacturing and electricity generation, and particular countries with especially large per capita emissions (like the USA), clearly bear significant responsibilities. What is needed is not the creative anger that might be vented against a corporate adversary (as in more localized environmental issues), but (inter-)national self-reflection towards adopting new lifestyles that save rather than waste energy. This will require wholesale transformations to existing industrial and social structures, and new policy priorities that support such changes.

In the end, global warming is an issue which requires that ordinary people exercise their imaginations. The current affects of global warming are only apparent through scientific reports that, for example, part of a glacier in the North Pole is beginning to melt, or unseasonal weather has become increasingly common in recent years, or that average temperatures are rising. In other words, it is not possible to 'know' about the affects of global warming except through the calculations and estimations of specialists and organizations such as the Intergovernmental Panel on Climate Change (IPCC), or from the information disseminated by public authorities and the mass media. Thus the nature of global environmental issues is such that they can only be tackled by processes in which public authorities take the lead. The basic structure of issues involved here are the same for national governments, municipal authorities, private enterprise, the mass media, academic researchers, and environmental groups—and all are dependent upon specialists.

However, an 'information war' is being waged in various forms over global environmental issues. Specialists disagree about the 'facts' of global warming as well as forecasted effects—and interested parties manipulate the information provided by specialists in their efforts to sway public opinion one way or the other. Thus while it is important to render the issues visible and comprehensible, efforts to do so often lead to over-simplification or distortion of the information, whether deliberately or not. Today, in the age of the internet and active environmental NGOs, the citizenry of developed countries are increasingly aware of the issues and can directly access masses of information through their home computers. They are not reliant on either governments or the mass media for their information. It is thus becoming increasingly possible for active citizens to criticize and expose attempts to manipulate information by governments, other interested parties (e.g. large corporations) and their specialists.

A brief examination of the history of the anti-nuclear movement in Japan sheds further light on the difficulties facing the environmental movement's efforts to redress the issues of global warming. From 1987 to 1989, a citizens' movement, the Anti-Nuclear Power Plant New Wave, erupted and spread across Japan, centered on housewives in large metropolitan areas (see chapter 8). This movement erupted onto the national stage, fueled by the April 1986 Chernobyl accident and the February 1988 Ikata Nuclear Power Plant's test to adjust output. The protest activity succeeded in stopping this test, but then faded away over the next few years. This case clearly supports the claim that, generally speaking, protest movements without a specific locale are difficult to maintain.

In the areas surrounding nuclear power plants, the issue at stake is clear and continuously present. The anti-nuclear residents' movement can neither run away nor relax its vigil. There are occasionally troubles or accidents in nuclear reactors, to which a movement must immediately respond. Most importantly, the nuclear power plant and its corporate owners provide tangible and direct targets for the movement.

In contrast, the anti-nuclear citizens' movements of the large metropolitan areas run study circles, collect signatures, liaise with the residents of areas where there are nuclear reactors, protest against electric companies, and write demands to local municipalities. But once they have done all of these things they often find it difficult to discover new issues or actions, and find themselves simply repeating the process. It is difficult to clearly identify any definite results of this movement. As people gradually lose faith in their political effectiveness, they grow weary, lose interest and the movement gradually loses its motive force and fades away. It should be clear that movements responding to global warming are similarly difficult to organize, and are inherently structured in such a way that it is difficult to build and maintain momentum.

Environmental NGOs and the Kyoto Conference on Global Warming

Despite these difficulties, Japanese and international environmental NGOs played a large role in the Kyoto Conference on the prevention of global warming in December 1997. This was the largest international conference ever held in Japan, with 9,850 individuals from 161 countries participating, including 3,865 people from 278 observer groups (not including news organizations) such as environmental NGOs and economic NGOs (Takeuchi 1998: 190). The roles played by these groups were varied and multi-faceted, including lobbying and/or criticizing the various national governments, collecting information, observing negotiations and witnessing agreements, engaging in dialogue with other NGOs, educating/informing the media and general public, and helping to shape public opinion both locally and internationally. It is probably safe to say that without the NGOs, there would not have been a Kyoto Protocol. At the same time, the conference provided a wide range of opportunities and venues for the representatives of the various NGOs to meet and exchange ideas with one another. Outside the conference, in Kyoto and its environs there were numerous associated events and meetings held every day. This was the best opportunity ever for those involved in the environmental movement in Japan, the mass media and the Japanese government, to experience first hand the power and enthusiasm of Japanese and international NGOs.[5]

The environmental movements are rapidly internationalizing, stimulated by the Earth Summit of 1992, annual conferences since 1995 by the member states of the United Nations Framework Convention on Climate Change, and most recently boosted by the Kyoto Conference on Global Warming. The Japanese Environment Agency's view of NGOs has significantly changed in the wake of the international acclaim for the Japanese umbrella-type NGO Kiko Forum (Climate Forum) as the host country NGO following the example of the Klima Forum at the first session known as COP-1, in Berlin, 1995. Thus Japan has entered an age of highly specialized and policy-oriented environmental NGOs.

4 Motivating and mobilizing the environmental movements

Factors that encourage or discourage participation

Volunteers and free riders

When walking around a new housing development in a city such as Sendai, one soon becomes aware of how much care is invested in the gardens of each and every house. Farmer's gardens also boast large beds of brilliantly colored flowers. Homeowners are passionate about keeping their gardens tidy. However, when the neighborhood association for the new housing area where I live organizes a voluntary working-bee to clean and tidy the public spaces, it is typically the same group of people from less than 20% of the households who participate.

While everyone desires an attractive and pleasant environment, relatively few are prepared to actively participate in the activities required to protect and maintain the environment. Most environmental protection groups suffer the twin headaches of a weak financial status and a small number of willing participants. Although citizens and NPOs have become involved in a wide variety of activities, their participation rates are often much worse than my neighborhood association's.

Why are people not more committed to voluntary environmental protection activities? What social conditions are necessary to promote greater environmental voluntarism among a populace? Social movement theories provide useful insights for considering these questions (cf. Hasegawa 1985b; 1990).

The first question—why are people not more committed to environmental or other voluntarism?—has come to be known as 'Olson's free rider problem' after the economist, Mancur Olson, who first noted it (Olson 1965). According to Olson, it is because people are egoists who want to enjoy the benefits of environmental or other voluntarism while avoiding the costs—i.e. they prefer to be 'free riders'.

Many efforts to promote participation in volunteer activities emphasize the importance and significance of information/education activities to increase public awareness. Olson, however, argues that this approach assumes a simple optimistic concept of 'pre-established harmony'. According to Olson, a person's awareness of a 'common interest' or 'common cause' is not in and of itself sufficient to motivate them to make an immediate contribution. Assume that a healthy and clean local environment is something that everyone in the area can enjoy equally and in common—that is, it has a characteristic of 'non-excludability' in which if the good is available to anyone, it is available to all. It is impossible to exclude anyone. Typical examples are clean air and public space in general. According to Olson's argument, people act to maximize their own interests. They will therefore avoid the required costs (time and effort), while ensuring that they enjoy 'their share' of the environmental benefits. Olson concludes that no one apart from the naively amiable will contribute unless special conditions are met.

From this perspective, Olson suggests that volunteer activities among self-interested people are dependant upon one of the following conditions: 1) The group is small enough that the 'free riders' are clearly observable, 2) the scenario separately offers 'selective incentives' beyond common interests corresponding to the level of each contribution, or 3) contribution is compulsory. In small mountain Japanese farming villages, close to 90% of the population votes in elections, and almost all households participate in local 'community clean ups'. This is due to a *de facto* compulsion. Who voted and who did not is readily apparent. However, compulsion is a contradiction to the concept of voluntarism, and so in the volunteer activities of social movements the third condition must be discounted. Also, the condition described in the small group scenario does not apply in considering situations where at least twenty or thirty volunteers participate.[1] Olson therefore stresses the importance of offering selective incentives, where participants are only 'rewarded' commensurable to their level of participation.

Generally speaking, in the business world, salary, position and promotions can function as rewards for individual contributions. In the case of environmental volunteers, this type of economic incentive cannot be provided in principle. A defining characteristic of volunteering is the absence of financial reward. Instead, less tangible rewards such as happiness, a sense of achievement, a sense of self-realization derived in or through the creation of a pleasant environment or fulfilling spiritual values must suffice for goal-oriented volunteers in environmental works. Some participants also experience a sense of fulfilling emotional values such as the joy of

meeting new people, forming relationships and working cooperatively
with others. The former have been classified as 'purposive incentives'
which include 'moral incentives and social redemption', the latter are
'solidary incentives' which are defined as 'the social and communal
sense of association with like-minded people' (Zald and McCarthy
1987: 81).

Incentives, contributions and resources

The time and labor costs shouldered by environmental volunteers
are a problem that cannot be ignored. Purposive incentives and
solidary incentives are therefore crucial in nurturing environmental
volunteers. In Japan, for the most part, middle-aged men cannot be
expected to contribute very much time, if any, because it is almost
wholly devoted to job commitments, and the predominant structure
of the Japanese workplace does not generally allow one to take time-
off in order to do volunteer work. Environmental volunteers have
therefore been largely limited to full-time housewives, retirees and
students. But middle-aged men are not wholly absent, either—those
who work relatively little overtime, who can leave work and return
home at a regular time each day, or can take paid leave, such as public
servants, teachers and the self-employed, are among the most willing
volunteers. It appears, then, that efforts to promote volunteer work
would be greatly enhanced by shorter working hours (ensuring a two
day weekend each week) and bringing the workplace closer to home
(thus reducing the time spent in commuting).

　In Japan to date, volunteers who are concerned with pollution and
other environmental issues have typically been found working with
residents' and citizens' groups. Frequently, a local businessman from
a highly reputed family assumes the role of leader, while a public
servant or teacher with, for example, trade union experience, acts
as the secretary, and the membership principally comprises retirees
and housewives. There is a clear pattern of volunteers in supporting
and planning positions often being male specialists in free-lance
professions such as high school and university teachers, lawyers,
doctors, certified public accountants and journalists etc.[2]

　These individuals not only have the time to become involved, but
also have a sense of mission (one example of a purposive incentive),
specialized knowledge, skills or leadership abilities. The ability
to access and mobilize various social resources is a vital factor
influencing volunteer activity. Resources can be divided into four

categories: people (human resources), material objects (economic/ material resources), symbols (information resources), and connections (relation resources). This is the perspective emphasized by the 'resource mobilization' theorists, a school of thought born in the late 1970s within studies of American social movements. As I have argued elsewhere, the Japanese residents' and citizens' movements can be fully explained in terms of resource mobilization theory (Hasegawa 1985b, 1990; cf. Shiobara ed. 1989 and Katagiri 1995).

From residents' and citizens' movements to environmental NPOs

The significance of being recognized as a legal entity

The passage of the Law to Promote Specified Non-profit Activities (commonly known as the NPO Law) in 1998 offers hope that in environmental issues as well as social welfare the number of volunteers in Japan who are organized around NPOs will grow. Before we proceed it is worth differentiating between NPO type environmental volunteerism and the activities of residents' and citizens' movements (as described in chapter 3). Salamon (1992) defines an NPO as satisfying six conditions: it must be a formal organization that is non-government, non-profit, autonomous, voluntary, and whose mission is publicly-oriented. Residents' and citizens' movements satisfy all but the first of these conditions—they are not formally legal organizations. Although there were cases where residents' or citizens' groups were formally organized, before the NPO law came into effect they could not be recognized as legal entities and were therefore only 'unofficial organizations'. This is the central characteristic that differentiates an NPO-style social movement from residents' and citizens' movements.

NPO-style voluntary organizations are in an ideal sense an extension of citizens' movements in contrast with residents' movements, as described in chapter 3. The residents' and citizens' movements have almost invariably responded to problems only after they have surfaced, responding after-the-fact and individually. They can therefore by characterized as defensive and critical protest movements against development-oriented corporations and governments. Once the issue of concern becomes less acute, it is not uncommon for such movements to discontinue their protest activities if they do not disband altogether. In that sense, many were single-issue, one-off movements. In other words, these were reactive movements in which a group of individuals furiously tried to stamp out the sparks that threatened themselves and their communities—brief explosions of protective activity triggered by an immediate sense of crisis that threatened the foundations of their everyday lives.

With juridical recognition as a legal entity, NPOs can retain full-time paid staff and establish a permanent board of management. It then becomes possible to tackle broader environmental problems on a continuing basis. It also becomes more likely to transform itself from a reactive movement to a proactive movement—acting to prevent environmental damage before it occurs. From there, it can be expected at the local, national and international levels to become a policy-oriented movement that develops and proposes alternative environmental policies.

The benefits of being a legally constituted NPO are most perceptible at the local level as, for example, in a citizens' group involved in recycling activities. Reactive single-issue groups derive little benefit from being recognized as a legal entity, particularly when their agenda directly conflicts with the plans or activities of large corporations and government agencies. In fact, such groups may find the bureaucratic requirements and processes required to be legally constituted unnecessarily onerous—incorporation requires submitting a list of founding members, an inventory of assets at the time of establishment, plans for future activities, annual financial budgets, final accounts of expenditures and revenues and so on. In contrast, an organization that is committed to recycling activity on a continual basis will enjoy significant benefits from juridical recognition in terms of things such as leasing office space, purchasing equipment (e.g. office equipment or a truck), and entering into commercial contracts with rubbish collecting companies etc. Legal status is important for consumer groups and those promoting tree-planting, renewable energy, or cleaning rivers and waterways. Citizens' groups that receive support from the Japan Junior Chamber, consumers' cooperatives, agricultural cooperatives and the local media, and that emphasize collaboration and joint efforts with local municipalities in Japan have begun applying for legal incorporation because it strengthens their networks and collaborative relationships. Thus there are two relatively distinct categories of NPO. First are grass-roots organizations with a non-specialist character. These are locally based and operate within a prefecture, city, town, village or other geographical area. These NPOs are an extension of environmental citizens' movements, and are easily accessible by local volunteers.

The second category of environmental NPOs are metropolitan based and of a much more specialist nature. They typically operate at a national level, but may possibly branch out to an international region or global level. At the national level, they may act as 'an alternative think tank' that remains independent of (and often in opposition to) government and private sector think tanks, or provides a kind of consultancy service for critical citizens against government and private sector organizations on environmental issues. It seems likely at this point in time that the Japanese branches of international environmental groups such as the World Wildlife Fund (WWF), Greenpeace, or Friends of the Earth will increasingly move in this direction. Environmental NPOs with Japanese origins that have achieved international success include the Kiko Forum (Climate Forum), the Citizens Alliance for Saving the Atmosphere and the Earth (CASA), and the Citizens' Nuclear Information Center (CNIC).[3]

The way to public affairs

We will have to wait and see whether environmental NPOs will be a mere passing trend in the current NPO boom or if it is possible for them to continue to grow and develop. As we have seen, residents' and citizens' movements tend to be concerned with issues that immediately affect them, seek urgent counter measures, and are motivated by compelling and direct purposive incentives. The sense of urgency comes from a fear that if they fail to act their local communities and local environment will be destroyed through development and pollution. At the level of protecting the local residential environment, the common good may be relatively concrete and tangible. Returning to our example from chapter 3, immediately after the Chernobyl nuclear accident in April 1986, the 'Anti-Nuclear Energy New Wave' movement was formed, primarily by housewives from large metropolitan areas. This movement was mainly motivated by a sense of real danger at the family dining table from contaminated European foodstuffs. However, from about 1990, this movement has been able to mobilize fewer and fewer people as memories of Chernobyl and corresponding fears of polluted foods have faded away (cf. chapter 8).

To the best of my knowledge, all of the anti-nuclear groups still active in Japan today—whether in opposition to the construction of new power plants or to the operational activities of existing power plants—are centered on women, and with the exception of those in the Tōkaimura and surrounding area that arose in response to the JCO accident, are located in the principal cities of prefectures that have nuclear power stations or related facilities (or neighbor such prefectures): that is, Sapporo, Hakodate, Hirosaki, Shizuoka, Hamamatsu, Toyama, Fukui, Kyoto, Osaka, and Matsuyama. In these areas there have been constant troubles associated with the nuclear institutions, and continuous debate about various plans to increase the size of existing facilities, introduce plutonium-thermal energy, and to construct new facilities. In these cases, the dangers and problems of nuclear energy have an immediate 'real-ness' that cannot be dismissed as someone else's problem.

However, as discussed in chapter 3, in the case of global environmental issues such as global warming there is at present no particular locale where the danger can be immediately sensed. In this instance, the collective good of concern—the environment—encompasses the entire globe. Moreover, estimated time frames for the affects of global warming to be felt vary from 50 to 100 years or more into the future. In other words, individuals will not feel its impact for several generations. There are thus no tangible purposive incentives, and the issues remain beyond the sphere of everyday cares.

As explained above, the silent majority of 'ordinary people' tends towards exclusive pursuit of private happiness. They will react swiftly to any danger to their own lifestyle—such as food contamination, or the destruction of their immediate residential environment by industrial pollution or high-speed transportation pollution—but they are slow to react to issues that lie outside of their everyday experience. The further removed a problem is from everyday life, the fewer the people who are likely to become active in a movement to redress it. Participation in resolving such issues is limited to perceptive people who can identify themselves on the basis of purposive incentives. Furthermore, generally the fewer the participants, the more likely the group is to be idealistic, and hence the more likely the movement is to become 'stoic' and principled, which serves to further distance the group from 'ordinary people'—a vicious cycle leading to even fewer participants.

Finding ways to break this vicious cycle and involve greater numbers of 'ordinary people' in the movement to redress global environmental problems is therefore a leading concern for Japan's citizens' movements and NPO activities.

Differences between volunteers in welfare and environmental movements

There are many similarities between welfare organizations and environmental movements—for example, both are dependent upon volunteer workers and financial donations. But there are also several important differences between welfare and environmental volunteering.

Welfare organizations have a clearly identified clientele, such as the elderly or disabled, who actively seek welfare services. Here, a deep, inspiring sense of satisfaction in having helped another may reward one's altruism. The purposive incentive is perceptible with the naked senses. One may feel a sense of social solidarity through establishing a connection with another person. Furthermore, although the extent of the government's responsibility is vigorously debated, in many cases there is no clearly identifiable perpetrator who is and can be held responsible for the harm incurred. The activities of welfare volunteers only minimally pose direct challenges to the existing social institutions and structures. It is therefore not uncommon for large corporations and government agencies to welcome, or even encourage their employees' involvement in voluntary welfare activities outside of working hours.

In contrast, and somewhat schematically, in working to redress environmental problems, it is generally not so easy to assess the extent to which one's contributions have alleviated the problem or to develop/maintain a true sense of personal effectiveness. Protecting the environment is an enormous task—inter-generational, global and never-ending—where individual and group efforts may seem to be miniscule, which can lead to an inner battle with hopelessness and ineffectiveness. Family and friends may ask 'Although it is a noble pursuit, does it really help?' Answering such questions is not always easy. The concrete contributions that can be made where results are more tangible, such as tree-planting, cleaning beaches and waterways, or recycling paper and clothing, is miniscule compared to the rate of forest destruction, the seemingly endless accumulation of cans and garbage on the beaches, and the incredible amount of paper and clothing that go to waste. Protecting beaches and waterways from the construction of new dams and landfills is not easy. It often seems as though an individual's participation does not have any impact.

Furthermore, in many cases of environmental pollution, there is a corporate or public entity that can be clearly identified as the perpetrator. Large-scale environmental destruction can almost always be attributed to the manufacturing activities of particular corporations. Even where the activities of ordinary consumers are directly implicated, such as the drink and fast food containers littering public spaces, the responsibility of manufacturing and retail companies is very large. Soft drink manufacturers, for example, sell canned drinks, continually advertise to encourage greater consumption of their products, install vending machines everywhere to increase distribution, and resist all efforts to introduce a deposit system or other recycling measures.

Environmental activists' efforts to combat the associated problems help to engender a critical awareness of modern society's mass production and mass consumption characteristics. In the process, individual companies and government agencies come under closer scrutiny and activists become increasingly critical of the wasteful consumerist lifestyle. The more passionate the environmental volunteer, the more critical he or she will be of society, businesses, governments and individuals. For government and corporate employers, the development of such attitudes among their employees and the community poses a very real threat, and environmental activism is therefore discouraged. Volunteer activists thus face isolation or mislabeling in the workplace, and friction with family members. Being a voluntary environmental activist entails shouldering risks—in society, the workplace, and the family.

Towards nurturing environmental volunteers

Paying and the significance of paying

As I have repeatedly argued, it is important to provide both purposive incentives and solidary incentives to nurture environmental volunteers. That is, it is necessary to create opportunities for volunteers to derive a sense of achievement, to ensure that voluntary environmentalists find their experience enjoyable and satisfying.

One form of participation that is widely available and has relatively low costs is to become a subscribing member of an environmental NPO by paying membership fees or making voluntary financial contributions. When Japan's NPO Law was enacted in 1998, there was extensive debate about whether preferential tax provisions should be made, but it was decided at the time to postpone such action. The law was revised in 2001, and measures were introduced to give donors preferential treatment. However, many restrictions were imposed and the system remains ineffective.[4]

In the USA, in contrast, section 501(C) (3) of the Internal Revenue Code provides special tax benefits for nonprofit organizations recognized as exempt from federal income tax. Donations to public interest groups (called charitable organizations) are fully deductible from the donor's taxable income. Even the cost of postage incurred in remitting donations is deductible. Eligible charitable organizations include grant-giving bodies, churches, private schools, private museums, zoos and so on. It is thus a much broader concept than the Japanese NPO. According to the Internal Revenue Service there were 870,000 recognized charitable organizations in the USA in 2001, an increase of 32.3% in five years.

The important point is that the provisions for reducing taxable income encourages citizens to provide financial support to charitable organizations according to their personal interests and beliefs. In 1998, 70.1% of US households donated an average US$1,075 per year (Independent Sector 2001).[5]

With their long history of social activism, social movement organizations in the USA are passionate about collecting funds, unlike Japanese social movements and citizens' activities which have for many years found fund raising to be the greatest obstacle in their activities—at least partly due to the absence of any systematic legal framework. It is nevertheless clearly possible and necessary to foster a belief that a financial donation is a significant contribution to redressing social issues.

In Japan in recent years, citizens' communal power plants are appearing in more and more areas, as citizens pool their funds to build solar or wind power installations, beginning with Miyazaki Prefecture, Shiga Prefecture and Hokkaidō (see chapter 10). The collective financial contributions of ordinary citizens have rendered possible the development and spread of an 'exemplary action' that challenges the power of the existing electric companies. This topic is pursued further in chapter 10, where I discuss how environmental NPOs can be developed into community businesses.

Environmental education

Environmental education is relatively weak in Japan, compared to the USA or Europe. 'Environmental education' refers to a practical program that teaches children and adults the significance of environmental protection, and the knowledge and techniques required, through school- or community-based education. Despite its relative weakness, though, there have been several innovative developments towards such programs in Japan. A prime example can be found in the work of Kada (1995). She attempts to tap into local memories of earlier life practices around Lake Biwa, beginning with individuals' and communal relationships with the waterfront. She employs various methods, such as getting people to talk about their recollections, organizing groups to observe fireflies, and exhibiting old photographs and artifacts such as millstones and Goemon baths, an ancient and special style of bathing. Her objective is to help local residents to remember local traditions of interaction with the environment, and to raise awareness of how and why these forms of interaction have changed.

Nurturing environmental NPOS

In simple terms, NPOs can be viewed as enterprises for the activities of citizens. To get an environmental NPO up and running requires influxes of money, people and information, as well as social arrangements that allow a new network to spread. In spite of its limitations, the NPO Law provides a legal framework for such enterprises. There is now a need for concrete ideas and schemes to proceed within this framework.

So-called 'intermediary organizations' and the NPO Support Centers have been created to provide support to NPOs, especially in nurturing people, collecting relevant information and developing/ maintaining networks. NPO Support Centers make policy recommendations to municipal and national governments on matters related to the operations of NPOs, and develop relationships with other intermediary, government bodies and private enterprises to exchange information, conduct lectures and study programs, train leaders and personnel, provide support and financial advice for creating and operating NPOs, and have developed NPO evaluation systems.[6] The number of NPO support groups in Japan has been increasing over recent years. According to the Japan NPO Center, as of late March 1999, there were a total of 40 such groups in Japan, of which 28 had been established principally in large metropolitan areas by the private sector (including groups affiliated with the Social Welfare Council), and the other 12 were created by municipalities. By the end of October 2002, this had increased to 75 groups, of which 42 were private sector and 33 were municipal.[7] In the ensuing three years the number has almost doubled, and the number established by municipalities has almost trebled. To date, however, almost all have been limited to a specific geographical area (a prefecture, city or town). There is a growing need for intermediary organizations that are not defined by geography but rather by categories of activity, such as environmentalism or social welfare services. Needless to say, NPOs are not panaceas. Salamon (1995), a well-known researcher of NPOs, proposes a concept of 'volunteer failure', with reference to established concepts such as 'government failure' and 'market failure'. He identifies four types of volunteer failure: the general shortage of resources, the gap between available resources and required resources (when resources are not allocated where they are most needed), benevolent patriotism, and amateurism (when the opinions of amateurs are given priority even where specialist advice is required). Such 'volunteer failures' have been all too common in Japan's environmental and citizens' movements. Overcoming these failures is therefore a central issue confronting Japan's environmental movements.

The self-appointed role of environmental NPOs requires them to maintain a fraught relationship with governments and private enterprises, critically challenging them while collaborating with them, strengthening their social surveillance functions, increasing their ability to investigate and expose problems, and to make policy proposals. Environmental NPOs are perhaps the only social organizations that can replace the declining labor unions and political parties, neighborhood associations and municipalities, and can call on a broader range of citizens to participate and contribute.

Citizens must be made aware that they hold the key to a radical shift in environmental policy, and must pioneer a diverse range of exemplary actions in various areas of modern life. Environmental NPOs and the volunteers who support them will be key players in the creation of a genuine civil society in Japan, and will thus determine the future shape of Japanese society in a very real sense.

5 Researching and developing environmental movements

Introduction

Of the various themes studied by environmental sociology, environmental movements are critically important. In fact the relationship between environmental movements and environmental sociology is far deeper than is indicated by the statement that the former is an object of research for the latter. In both Japan and the USA, environmental sociology has been stimulated and nurtured by the environmental movements. Why was this so? What impact did the have on environmental sociology? What new insights can sociology offer for understanding environmental movements?

In this chapter, I begin with an examination of the role played by the environmental movements and the significance of social movement research in the formation and development of environmental sociology. Then I discuss the merits of three approaches to analyzing environmental movements—cultural framing, structures of political opportunity, and mobilizing structure. Finally I analyze the characteristics of the environmental movements as new social movements to ascertain what social functions environmental movements fulfill.

Environmental sociology and the environmental movements

Environmental sociology and the environmental movements in Japan

In general, wherever environmental issues are recognized, some sort of social movements will emerge. Indeed, one might go so far as to say that where there is an environmental problem there will be an environmental movement. The absence of a social movement to redress a grave environmental problem should therefore be seen as evidence of restrictions on the media's freedom to report and the public's right to protest. In other words, in such cases it is political/social factors that prevent a social movement from developing.

Most of the articles and survey reports written by environmental sociologists discuss an environmental movement that has been triggered by a particular environmental problem. Where the particular movement itself or social movement theory is not clearly identified as a central topic for the research/paper, it is very rare that there is no mention at all of related movements. For instance, Hasumi's (1965) pioneering analysis of problems surrounding a new industrial complex at Yokkaichi was part of an empirical research project to examine issues of development. Pollution was treated as a negative aspect of industrial development that can be damaging to human lives. In his analysis of pollution, however, Hasumi describes how the residents' movement responded to the environmental problem of the Yokkaichi industrial complex, including detailed analyses of the relationship between the movement and the political structure, and outlines the responses of political parties, labor unions and local government.

As mentioned in chapter 1, the first paper that could be un-ambiguously called a work in 'environmental sociology' published in an academic sociological journal in Japan is a comparative analysis of two outbreaks of Minamata disease (caused by mercury poisoning) in Kumamoto and Niigata (Iijima 1970b). Iijima reveals that the actions taken by patients and their supporters in the two cases were different, as were the processes of the two movements. She argues that these differences were determined by the local community structures, relationships between the local communities and the polluting companies, the social class of the affected persons in each place, and the chronological order of the incidents. Iijima's study is a classical example of the sociological approach. She says, 'In the history of environmental sociology research until the 1970s, research by sociologists into pollution and environmental problems generally employed the methods of region and community studies and rural sociology, focusing on the analysis of protest movements and residents' movements' (Iijima 1993: 220).

In 1985, Funabashi, I and two other scholars published the results of our research into bullet train pollution and associated problems in the construction of the bullet train lines. A central focus of my part of this research was analyzing the residents' movement and the court cases. Other central issues were the characteristics of the problem from the perspective of the actors—both beneficiaries and victims—elucidating the mechanisms which enabled the (then) Japan National Railways to ignore the problem, and a critical analysis of the JNR's perception of 'public works' and the usefulness of bullet trains (see chapter 7).

Although the problems arising from large-scale development and waste disposal remain central concerns of environmental sociology today, analysis of the logic, organization, and developmental processes of protest movements are a central pillar of sociological analysis in Japan (Funabashi et al. eds, 1998; Ukai 1992). Research into efforts to promote organic agriculture, for example, by sociologists such as Aoki, Matsumura and Masugata have focused on the movement to develop cooperative links between urban consumers and rural producers and the challenges faced by farmers who attempt to adopt organic agricultural practices (Kokumin Seikatsu Sentä ed. 1981; Aoki and Matsumura eds, 1991; Masugata 1995).

The life-environmentalism developed by Torigoe and others have provided detailed studies on the problems facing the communities surrounding the Lake Biwa, Nagara River and Toga River areas. Although they have been criticized for not dealing with the acute social disputes and severe fissures in Japanese society, in my view they have provided strong insights for scholars of social movement theory. They discussed various efforts to protect the environment with a focus on the local residents' participation, including historical analyses of traditional life practices around lakes and waterways. These analyses were then used in local educational programs— educating local residents' about their traditional life-practices while promoting consciousness of the affects of environmental changes— both on the individual human body and the in the culture of their communities (Torigoe and Kada eds, 1984; Kada 1995; Adachi 1999; Torigoe 1997). As described in chapter 4, in Kada's practice—Kada was at the time a staff member at the Biwako Museum—life-environmentalism aimed to help local residents to remember local traditions of interaction with the environment, and to raise awareness of how and why these forms of interaction have changed. This was, I believe, a unique contribution to the theory and practices of social movements, especially the environmental education movement.

The local communal management system developed by Nakata and others focuses on the structure of self-governing local residents' organizations in order to achieve some form of local co-management. The research emphasized the composition of the residents' and neighborhood organizations and their function in creating a community (Nakata 1993). Residents' self-government and the creation of communities are central themes in this argument.

Similarly, sociological research into the issues surrounding the protection of cultural heritage (streets and buildings) typically focuses on the movements involved and their internal processes (Yoshikane 1996; Horikawa 1998).

In Japanese environmental sociology, Umino's quantitative analyses of social dilemmas use game theory as a framework. When applied to waste-producing behavior, therefore Umino and his followers assume the existence of equal and homogenous actors. Relations of power and domination are ignored in order to simplify the model. As a result, their framework is inadequate for analyzing environmental movements (Umino 1991; Nakano et al. 1996).

In short, the analysis of social movements plays a large role in each of the schools of thought in Japanese environmental sociology except for Umino's quantitative method. Further evidence of the importance of the environmental movements for Japanese environmental sociology can be found in the structure of volume 12 of *Sociology in Japan* (*Kōza shakaigaku*), entitled *Environmental sociology* (*Kankyō*) (Funabashi and Iijima eds, 1998). This volume consists of seven chapters, of which two are general overviews by the editors, and the remaining five present analyses of individual topics. Three of these five include the word 'movement' in their titles ('The organic agricultural movement', 'The protection of historical streets and houses movement', and 'The environmental movement in Japan and the USA').

Iijima divides environmental sociology into four research areas— environmental issues, environmental symbiosis, environmental behavior and environmental consciousness/culture—and notes that in Japan 'the overwhelming majority of research consisted of "sociological research of environmental issues" and "sociological research of environmental behavior"' (Iijima 1998b: 2–3). Much has been written about issues where these two fields intersect—e.g. the environmental movement as a victims' movement or victims' support movement, and the environmental protection movement. By comparison, sociological aspects of everyday activities that aim to reduce the human impact on the environment, such as recycling activities, or into the behavior/beliefs of perpetrators remain relatively unexamined and are areas where we look forward to future research.

Many of Japan's outstanding environmental sociologists entered this field via the sociological study of social movements. While the greatest numbers of the founding generation of environmental sociologists in Japan came from rural sociology and regional and community studies, the sociology of social movements was the third largest source of these personnel. My own work is primarily in areas where environmental sociology and the sociology of social movements intersect. Kamon Nitagai, Akihiko Takada, Ryōichi Terada, Shinji Katagiri and younger scholars such as Reiko Seki, Yūko Takubo (Hirabayashi), Won-Cheol Sung, and Hideo Nakazawa have also made important contributions in this area. The primary objects of research here are residents' movements, citizens' movements, NGOs and NPOs as they relate to environmental problems. Especially in Japan, these movements are typical of post-1960s protest activities, with strong similarities to labor, peace, student, and feminist movements. Environmental sociology in Japan has been strongly shaped by the sociology of social movements, as well as rural sociology and the sociology of local communities.

The strategy of the vantage point of movement theory in environmental sociology

The situation is almost exactly the same in the USA. According to Humphrey and Buttel, 'The tremendous amount of attention given by social scientists to the environmental movement can be gauged by the fact that roughly 350 studies of the movement had been written by 1977' (Humphrey and Buttel 1982: 24). The environmental movement that arose and gathered momentum in the USA in the late 1960s and early 1970s can be seen as a 'midwife' for the birth of environmental sociology in the USA. At almost the same time, interest in environmental problems grew in Japan and elsewhere around the world. The first Earth Day on 22 April 1970—a nationwide success in the USA—and the United Nations Conference on the Human Environment held in Stockholm in 1972 are indicative of this growing social awareness.

The surge in the environmental movements in the late 1960s provides the background to the emergence of environmental sociology. Until environmental sociology proclaimed itself to be an independent discipline in the early to mid 1970s in the USA, much of the embryonic research in this field was analyses of environmental movements from the perspective of the theories of social movements. Another 'midwife' in the birth of environmental sociology was research by rural sociologists into efforts by farmers to protect their natural resources. In other words environmental sociology in the

USA was born of two parents—the sociology of social movements and rural sociology—against a background of growing social interest in the environment. In 1976 the Section on Environmental Sociology was established within the American Sociological Association. However, while many prominent social movement theorists in Japan are environmental sociologists, this is far less common in the USA, where only a few authors—such as Morrison (1973) and Broadbent (1998)—work in areas where environmental sociology and the sociology of social movements intersect. This is at least in part because in the USA, the civil rights and women's movements have carried far more weight than elsewhere, and the impact of the environmental movements on the sociology of social movements has been correspondingly limited.

In the USA, Japan and Europe, active environmental movements had a significant influence on the birth and development of environmental sociology. Environmental ethics was born in the USA against a similar background—it emerged from the 'movement for environmental ethics' and became an independent research field.

The reasons for this close relationship between environmental movements and environmental sociology must seem obvious to anyone with even a little bit of sociological training. The growing concerns over environmental problems in the late 1960s were not simply due to the existence of environmental damage or risk, but also to the discovery, exposure and framing of such damage by environmental activists. Rather than a direct relationship between environmental problems and environmental sociology, then, we find that this relationship is everywhere mediated by the environmental movement. In this sense, the environmental movement constructed discourses of environmental problems. Indeed, disclosing environmental problems is perhaps the most important function of the environmental movement.[1]

Funabashi (2001) identifies three relatively distinct areas in the sociological research of environmental problems: identifying causes or perpetrators, victims, and solutions. As argued elsewhere, I believe understanding is better served through a four-fold distinction: perpetrators, victims, movements, and policies.[2] This four-fold distinction gives due recognition to the importance of the role played by the environmental movement in identifying problems and developing/implementing solutions. At the same time it highlights the importance of social movement theories in environmental sociology. Funabashi's three-fold scheme tends to occlude the significance of the mediating role that environmental movements play between victims and policy-makers.

The identity of environmental movements and environmental sociology

Although the preceding argument might seem to suggest that social problems are generally exposed by social movements, or that social movements typically provide the keys to resolving social problems, this is not necessarily so. Whilst it is true that the analysis of environmental problems and the analysis of environmental movements overlap to a great extent, this is not always the case with other social problems such as family troubles, agricultural problems and urban problems.

For instance, the sociological analysis of rural village issues does not always overlap with the sociology of agricultural movements or farmers' movements. When the organic farming movement, a topic which has a deep relation with environmental problems, is the focus of examination, the organic agricultural movement is typically a central concern (as noted above). But it is rare to see movement theory used when the focus is on the various problems facing rural agricultural communities today, such as the price of rice, adjustments to rice production, outside (off-farm) employment, the organization of the agricultural cooperative, a shortage of wives, the maintenance of rural communities and the disinterest of children in succeeding as the family head, increasing difficulty in finding people to take over farms as older farmers retire. This is because, with the exception of the organic agriculture movement, the movement to create and maintain communities, and the struggle by organizations such as the agricultural cooperative to increase the price paid for rice, phenomena linking rural communities to movements are not common.

At the present stage, it is difficult for the social issues related to the family to become the focus of social movements, with the exception of feminist movements that focus on the rights of women in the family. With regard to urban issues, it is easy for problems of the urban environment, urban renewal, and the rights of minorities to become the focus of movements. However, there is only limited overlapping between urban issues and social movements related to the city. It is uncommon for social movement analysis to play any significant role in rural sociology, family sociology or urban sociology. Why, then, does the sociology of the environmental movements play such an important role in the sociology of environmental issues?

The first point that can be made from the examples cited above is that it is uncommon for a problem to become the focus of a movement if and when there are routes open for the individual to solve the problem by him- or herself, or when there are existing social institutions that might solve the problem, such as family ties, local communities, residents' networks, or local government. Because functioning channels exist to tackle many of the problems of rural farming communities, family issues in rural communities, and so on, new social movements are only rarely organized. Putting this the other way around, then, the problems of organic agriculture, women's rights, the historical environment of cities, and minorities became the focus of new social movements because of the absence of functioning channels to resolve them.

In the case of most environmental problems, social movements play a strategic role in many cases because existing routes to tackle and solve problems were found to be inappropriate or inadequate. In cases of continuing or foreseeable environmental destruction, where there is a government agency actively attempting to respond to the problem, the contribution that an environmental movement can make remains limited. The true *raison d'être* of the environmental movement arises when faced with dysfunctional governance or political channels.

However the analysis of social movements and social movement theory have little presence in the environmental research conducted in other social science fields. Rather schematically, environmental law focuses on questions of whether rights have been violated or not, and how to restore violated rights. Based on the assumption of positive law, the main focus is on the victim and policy responses. Close analysis of the environmental law specialists' fieldwork reports, such as Awaji Takehisa, and the environmental economists Miyamoto Ken'ichi and Teranishi Shun'ichi, reveals that their positions entail an implied or *de facto* movement theory perspective or sympathy for movements. However, generally speaking, environmental law and environmental economics researchers are far less interested in the insights of movement theory and the environmental movements than environmental sociology. Indeed, in the textbooks and introductory works of environmental law and environmental economics, there is almost no mention of the environmental movements.

The perspectives developed by social movement theory thus give environmental sociology a unique identity distinct from environmental economics or environmental law. To date, the main strengths of Japanese environmental sociology have been its capacity to explain the victims' perspective, the social structures in the relationship between perpetrators and victims, and its theoretical understanding of the residents' and citizens' movements. Sociological analysis is perhaps uniquely suited to studying the phenomena of environmental movements from the perspective of social action, with its emphasis on fieldwork and the function of local actors.

The environmental movement as an analytical perspective

The triangular structure of social movement analysis

Environmental movements are special types of social movements that arise and are active in response to environmental problems. The analytical frameworks and concepts of social movement theory can therefore be used in principle without qualification.

As discussed in chapter 3, I define social movements as 'collective actions oriented towards change that are based on grievances and discontent with the status quo or an anticipated state of affairs' (Hasegawa 1993a: 147). The environmental movements can be defined as collective action oriented towards change that is motivated by grievances and discontent with the (present or foreseeable) condition of the environment. I will now explore the significance of this definition through an examination of characteristics that are common to almost all social movements and those that are unique to the context of environmental problems.

The crucial terms in this definition are: discontent, orientation for change, and collective action. 'Discontent' refers to the motivations for people to participate in a movement. In the history of social movement theories, the 'collective behavior approach' of Geschwender (1968), Turner and Killian (1972), and Smelser (1963) attempts to explain collective behavior, including social movements, through dissatisfaction at the micro level of individual actors. Geschwender's 'relative deprivation theory' is a typical example of this approach.

For discontent with environmental conditions to mobilize a social movement, it cannot be simply an individual matter, but rather a collective sense of deprivation, discontent or crisis must be experienced by some of the members of the affected community. Early resource mobilization theorists argued that this type of theory of collective behavior was inadequate for explaining the civil rights movement, noting that discontent in this case had continued for generations and should therefore be seen as a constant, and thus a secondary factor in motivating members. However, in the environmental movements, the sense of crisis, grievances and discontent about environmental damage and future risks are new experiences directly resulting from perceived damage, the clarification of cause and effect relationships, or the damage anticipated to arise from planned future developments. In other words, the environmental problems that have come to light in advanced industrial societies produce sudden explosions of collective deprivation. They are therefore primary factors in the development of the environmental movements.

The psychological and cultural factors involved in giving meaning to an unsatisfactory situation emphasized by social constructivism and cultural framing theory should not be ignored. However, as Dunlap (2003) argues, in environmental problems, the objective/structural basis for collective deprivation must not be overlooked because environmental problems cannot be resolved without rectifying the material conditions of collective deprivation or dissatisfaction.

The second critical term, 'orientation for change', refers to a movement's aims and values. Such an orientation as well as an organizational structure is what distinguishes a social movement from other forms of collective behavior such as panics, rumors, and mass hysteria.

A movement's orientation for change refers to a will to change the social structure—in whole or in part—such as the laws, political system, or the economy, or an even broader objective of changing social norms and values to focus beyond the personal desires and benefits of its members. In cases where the orientation is towards changing the social structures or systems, macro-level arguments are required. Marxism and the 'new social movement' theory are sociological perspectives that focus on a movement's orientation for change. 'New social movements' are characterized by Melucci (1984; 1989; 1996) and Cohen (1985) as 'self-limiting radicalism', referring to their orientation towards changing certain existing social structures—for example, defending the autonomy of civil society and enlarging the public space and the public sphere—

while accepting the basic structures of the market economy and parliamentary democracy, unlike, say, traditional Marxism which would change it all.

Many environmental movements share this self-limiting radical characteristic. In fact, in comparison to other new social movements such as feminism, peace and minority-rights movements, the environmental movement is especially seeking to minimize and, where possible, eliminate environmental risks, and is strongly oriented towards concrete and direct solutions to environmental problems.[3]

The third pivotal term, 'collective action' should already be clear, for as mentioned above, a social movement only exists when grievances and discontent are experienced, and orientation for change is adopted, by a 'collective actor'. Some sociologists of social movements focus on identifying various forms of collective action in social movements, especially in mezzo-level collectivities and formal organizations. Resource mobilization theory, for example, focuses on understanding movements at this level.

It is extremely difficult to resolve environmental issues through individual action. Strategies that incorporate collective, organized action against the perpetrators of environmental destruction and use the mass media to this end are indispensable. Following the Earth Summit of 1992, the role played by environmental NGOs and NPOs came to the attention of an increasingly international public.[4] In recent years organized collective action has come to have ever increasing social significance.

On the basis of these three dimensions—grievances and discontent, orientation for change, and collective action—the environmental movement is clearly a social movement.

In the 1990s, the primary trends in social movement research—collective behavior, new social movement, and resource mobilization theories—all moved towards more comprehensive explanations. Here, the key terms are cultural framing, the structures of political opportunity, and mobilizing structures (McAdam et al. eds, 1996; McAdam et al. 2001).

Cultural framing aims to define the shared world-view of the participants that justifies collective action and social movements. This 'frame' provides the motivation to participate—both an 'image of the world' and a 'self-image' of the movement. 'Framing' is the conscious and strategic process by which a frame is formed. 'Cultural framing' is a non-static process that mediates dissatisfaction and orientation for change. Influenced by Goffman and symbolic interactionism, Snow and others criticized the resource mobilization theory's explanation of the participants' motivation for being insufficient. They focus, instead, on the interaction between an organization and its members, and the processes through which the interests, values and beliefs of the individual participants on the one hand and the aims and activities of the social movement organization on the other are adjusted and brought into alignment. Framing analysis then provides the analytical framework for explaining the motivation underlying participation and commitment (Snow et al. 1986; Benford and Snow 2000; Hongö 2002).

The 'structure of political opportunities' is an alternative explanatory framework that integrates the political sociological version of the resource mobilization approach developed in America as articulated by Tilly, Obershall, McAdam, and Tarrow, with the inter-state comparative analysis of social movements developed by Kitschelt (1986) and Kriesi et al. (1995). Each author has a different view about which conditions to focus on but, as McAdam (1996: 27) explains, these can basically be synthesized into the following four categories: '1. the relative openness or closure of the institutionalized political system, 2. the stability or instability of that broad set of elite alignments that typically undergird a polity, 3. the presence or absence of elite allies and 4. the state's capacity and propensity for repression'.

'Mobilizing structures' refer to a set of variables at the resource level as defined by the economical sociological version of the resource mobilization approach. Here, the focus is on what resources

Figure 5.1 The triangular structure of social movement analysis

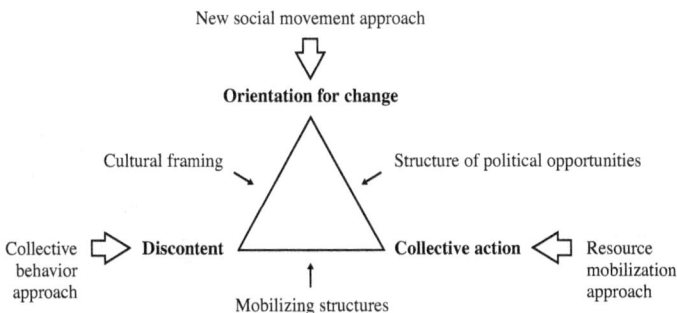

can be mobilized and under what conditions (Zald and McCarthy 1987).

If these are overlaid upon the three elements of social movements—discontent and orientation for change, and collective action—the result can be schematized as shown in Figure 5.1. Cultural framing mediates between discontent, orientation for change, the structure of political opportunity mediates between orientation for change and collective action, and mobilizing structures mediate between collective action and discontent.

Environmental movements as new social movements

There are many types of environmental movements in Japan. Distinguishing the major environmental movements by their goals, there are (1) anti-pollution protest movements demanding the prevention of further pollution and relief for victims, (2) nature protection movements that oppose large-scale industrial development projects and other threats to the natural environment, (3) environmental education movements focused on developing environmental consciousness, and (4) amenities improvement movements that seeks environmental improvements in residential districts, including the provision of parks and cleaning rivers and streams. There are also (5) organic agricultural movements, including efforts to directly link producers and consumers, and (6) movements to protect cultural heritage, historical streets and houses. In the 1960s and 1970s, the environmental movement consisted almost entirely of the first two types but has since become increasingly diverse. The anti-pollution movement includes the anti-tobacco movement's campaigns for a smoke-free environment and the movements opposing the use of nuclear reactors for generating electricity.

It is difficult to discuss all of these in general terms, but within the theoretical framework of new social movements, the environmental movements have the following characteristics. First, in contrast to the labor movement defined by the class struggle or a political movement that seeks wholesale transformation of the social establishment, the environmental movement is a typical example of a 'new social movement'. I will explain this by addressing each of Offe's (1985) four defining characteristics of a new social movement: (1) actor, (2) issue, (3) value-orientation, and (4) mode of action.

These are very similar to Touraine's (1985) (1) identity, (2) focus of conflict (*enjeu* in French), (3) cultural orientation, and (4) action.

Actors

There is a wide variety of actors in the environmental movements. Typical examples include farmers and fishermen, local community residents, general citizens, and individuals with specialist or higher education. Globally, women have played an important role. Environmental movements can be divided into residents' movements, and citizens' movements depending on the characteristics of the principal actors. In simple terms, residents' movements aim to directly defend the lives and interests of the residents of local communities, while citizens' movements are more ideal-oriented attempts to defend universal values by conscientious citizens (see chapter 3). Many of the environmental movements in Japan to date can be characterized as residents' movements. Citizens' type environmental movements such as those found in the USA and Europe have, in Japan, tended to be limited to activities in large metropolitan areas.

Issues

While the issues confronting the labor movement have revolved around the place of production, the workplace, the issues facing the environmental movement mainly revolve around the place of consumption, the sphere of everyday life. The term 'environmental movement' refers to an issue.

Among the new social movements, the environmental movement can be distinguished by its risk-avoidance. It is a movement that seeks to protect everyday life from environmental risks, by redressing environmental hazards that have become apparent and averting those that have been predicted. It does not seek to reclaim rights as do feminist and minority-rights movements, nor to reallocate material goods as do labor and welfare movements.

Value-orientation

Collectively, the environmental movement amounts to a wholesale critique of a civilization driven to conquer nature through technological 'progress', a civilization that prioritizes economic growth above all other values, and the accompanying mass-production, mass-consumption and mass-waste culture. The environmental movement opposes the technocrats who perpetuate this system, and promotes the value of environmentalism, which emphasizes the ecosystem, a 'symbiosis with nature', and a sustainable society. The environmental movement is frequently called an ecological movement, a terminology used when the movement is defined by an ecological view of the world or environmentalism. In this sense the environmental movement is strongly characterized as a value-oriented movement.

Mode of action
Environmental movements typically promote self-determination, self-expressionism and self-limiting radicalism. In opposition to technocrats, emphasis is given to forms of protest and expressions of dissent through acts of civil disobedience such as 'human chains' and 'die-ins'. The 'human chain' is formed by people holding hands along a protest route or surrounding a targeted facility. A 'die-in' involves protestors lying in the streets or a public square to symbolize the victims of an environmental disaster. Environmental movements are frequently in conflict with technocrats, private enterprise and government. Generally there is a corporation or other entity that can be clearly identified as the perpetrator, for environmental destruction downstream is almost always closely linked to the production activities of private enterprises upstream (as described in chapter 2). Commitment to an environmental movement requires a critical attitude towards modern consumer society, the companies that underpin and perpetuate this society, and the governments that have failed to introduce appropriate regulations. This kind of critical perspective often leads to a critical reflection on and greater awareness of one's own resource-intensive lifestyle. The environmental movement can also be characterized as a movement of dissent and contention, and thus the types of collaborations with government and private enterprise that are often found, for example, in the welfare movement, are relatively difficult to sustain. The environmental movements clearly embody the characteristics of new social movements.

The Social Function of the Environmental Movements

What social significance can be attributed to or expected from the environmental movements? Based on surveys investigating the problems of bullet train pollution and construction (Funabashi et al.

1985; Funabashi et al. 1988), we can discern the following general
social functions of environmental movements (see chapter 7).[5]

First, environmental movements function to publicly disclose
environmental issues as social problems, exposing the nature, site
and source of problems to public view. Here, the existence or
probability of environmental problems is not necessarily a social
problem until they have been defined as such. Environmental activists frequently identify perpetrator-victim (or causal) relationships,
identify the responsible parties, and assess the urgency of responding.
In this sense, environmental issues are socially constructed. In
some cases the local residents' understandings have highlighted
problems that the experts had been unable to identify.

The anti-bullet train pollution movements and the anti-bullet train
line construction movements of the 1970s are prime examples of
environmental movements in this role. Without these movements, the
'dark side' of the bullet train noise and vibration would not have been
regarded as a grave issue by the JNR, Ministry of Transportation,
local governments, media, passengers or general public.

Second, there is a problem-solving function at the level of
individual and concrete environmental problems. Here, once an
environmental movement establishes that an issue is a social problem,
responsible parties (enterprises and governments) are under pressure
to respond, and the environmental problem is resolved to some extent
and future harm is prevented or reduced.

In the anti-bullet train pollution movement in Nagoya, although
the central concrete objective of making the trains travel more slowly
was only realized for nine years, and only through the supportive
actions of the two major labor unions representing the employees of
the JNR, there were other concrete results. Policies were introduced
to reduce noise and vibration at their source, and measures such as
improving sound-barrier walls along the route were implemented, as
were other measures to compensate for the relocation of housing in
especially damaged cases.

Third, environmental movements often provide the impetus
required to reform existing social systems and norms. This can be
achieved through various means and at various levels, for example:
(1) successfully lobbying to introduce new or change existing
administrative and legislative mechanisms, (2) establishing legal
precedents through judicial decisions in court cases, (3) expanding
and strengthening civil rights, such as promoting the new concept
of an 'environmental right' (see chapter 7 note 8) , (4) promoting

general public awareness of environmental problems and their impact, and (5) demanding appropriate compensation for the victims of environmental problems.

Granted, concrete improvements and policy responses such as those mentioned often require that action be taken by the enterprises or governments concerned. But such actions are often taken as part of substantial changes in systems and norms that were stimulated by environmental activists. As a rule, the more serious an environmental problem is, the more vigorous the activism will be, and hence the more likely that widespread changes will be introduced.

In the case of bullet train pollution: (1) the government announced new environmental standards to reduce bullet train noise pollution, and introduced a new system for responding to bullet train noise problems; (2) in the Nagoya bullet train trial, although the court rejected the request for a suspension order, it recognized that the JNR was at fault and awarded damages; (3) the expansion and strengthening of civil rights, (4) the severity of bullet train noise pollution received great public exposure through the mass media and general awareness of the issue by ordinary citizens was quite high. Furthermore, there was a growing movement to challenge the 'public work' of the bullet train, as the social problems were disclosed and responses demanded through litigation. Subsequent environmental rights law suits in which the JNR or Minister of Transportation were the defendants included the case brought by residents in the southern part of Saitama Prefecture, including what was then Urawa City, to prevent construction of the Tōhoku and Jōetsu bullet train lines, an almost identical case brought by residents in Kita ward, northern Tokyo, and the anti-smoking litigation that demanded the wider adoption and increased number of non-smoking cars for bullet trains. All of these cases were lost in court, but the anti-smoking movement nevertheless achieved their goals. Today more than 70% of bullet train cars are non-smoking cars, in contrast to the single non-smoking car for non reserve seats per bullet train service when the law suit began in 1980.

Fourth, environmental movements serve to prevent or reduce further environmental damage. Reforms to systems and norms contribute not only to resolving the specific problems that the movement was originally concerned with, but also has a flow-on effect, restraining the emergence of similar environmental problems elsewhere.

Again, the problem of bullet train pollution and the anti-pollution movement in Nagoya provides a good example. Here the creation of new standards and systems contributed to an improvement in noise and vibration controls on the Tōhoku and Jōetsu lines, as well as technical improvements to the cars on the Tōkaidō and San'yō lines. The same social functions can be seen in Japan's four major pollution cases and the anti-pollution movements associated with them and in the USA's infamous Love Canal incident (1978) where dangerous industrial wastes contaminated the soil and the victims organized a movement to fight for justice. The strong and growing opposition to Japan's four major pollution cases led to new environmental laws being enacted by Japan's so-called 'pollution Diet' of 1970 and the establishment of the Japanese Environmental Agency in 1971. The Love Canal movement—centered on the housewives and other victims of this incident—led to the Superfund Law (1980) requiring those who had dumped the waste to restore the contaminated soil to its original condition (based on the principle of no-fault liability).

Again we must acknowledge that the environmental movements were not solely responsible for the new standards, regulations and legislation in these cases. In fact, each was the product of complicated political and social processes that unfolded over a considerable time and involved environmental movements, private enterprises, governments, the judiciary and others. Nevertheless, it is safe to say that none of these changes would have been affected were it not for the presence and activism of the environmental movements. In this sense, the environmental movements are indispensable constituents in the democratic processes of a civil society that must protect the environment and avoid environmental risks.

6 The environmental movements and policy research

Environmental sociology and policy research

Introduction

As a result of the historically closed-nature of the policy-making processes in Japan, the environmental movements and environmental policy-makers have long been seen as opposing camps. This unhappy state of affairs continued for many years, with research articles that discussed environmental movements rarely addressing environmental policy, and articles that discussed environmental policy barely mentioning the environmental movements. Under these circumstances a disciplinary demarcation developed whereby environmental sociologists almost exclusively discussed environmental movements, while scholars of environmental law or economics debated environmental policy.

The environmental movements have, however, been highly critical of policy and policy-makers, and have made slow progress in shaping policy responses to the environment. More recently, in Japan and elsewhere, environmental NGOs/NPOs have become increasingly policy-oriented, proposing and negotiating policies to redress environmental problems. The numbers and diversity of NGO/NPO participants at international conferences such as the Kyoto Conference to Prevent Global Warming provides evidence of this change. Following the Earth Summit in Rio de Janeiro in 1992, partnerships between government policy-makers and NGOs has been reassessed and encouraged at the United Nations, international conferences, and by local governments. The relationship between the environmental movements and authorities in charge of policy over the past decade or so has changed dramatically in Japan as well. In the roughly four years from December 1998, when organizations in Japan were first permitted to apply for recognition as non-profit corporate bodies, 14,657 organizations have been recognized (as of the end of December 2003, see Figure 13.1). In the field of the environment, a large number of environmental NPOs have come into existence, and the formal organization of the environmental movements is proceeding rapidly. If we are to build a sustainable society and transform current society into a recycle-based society, policy collaboration and partnership between the environmental movements and government authorities is essential. But is it possible, and if so, how? Sociology offers numerous analytic tools for answering this question, including methods to explicate the ideals and values that drive the environ-mental movements, and to identify the structural barriers of the existing social and political systems.

In this sense, understanding the dynamic relationships between the environmental movements and government policy is a very current and critical issue, both for environmental sociologists and their students, and for the environmental movements themselves.

The status of policy research

The importance of the environmental movements in environmental sociology was discussed in chapter 5. The sociological approach to environmental problems includes discourses on perpetrators, victims, movements and policies. One of the most important roles played by the environmental movements is mediating between the perpetrators, victims and policy-makers about particular environmental problems. The capacity to analytically delineate the multiple roles that are performed by the environmental movements provide environmental sociology with its unique identity.

And yet, while the environmental movements have played a defining role in the development of environmental sociology, environmental sociologists have rarely mentioned environmental policy.[1] Mr. Suda Harumi has been involved for many years with environmental movements and the citizens' legislation movement. He was invited as a panelist to the first joint symposium of the Japanese Association for Environmental Sociology, the Society for Environmental Economics and Policy Studies, the Japanese Association for Environmental Law and Policy and others, where he criticized the status of environmental sociology. He said 'I have no idea what use sociology is in solving the problems that are actually occurring. I have come to believe that it does nothing but categorize and follow up' (Awaji et al. eds, 2001: 74). Environmental sociologists must address this critique.[2]

Needless to say, the context of environmental problems includes certain developmental and environmental policies, industrial, transport, energy, waste, urban and agricultural policies and their practices. Therefore, research into environmental problems, whether pollution studies or research of developmental problems, organic agriculture, the problem of waste, the protection of historical streets and houses, or sociological research, almost without exception adopts a standpoint that is critical of policy.

However, to date, critical policy debate from environmental sociology has generally focused on the causes of or structural factors behind environment destruction crises. Relatively speaking, critical analysis of environmental policies themselves has been insufficient, as has research into the policy-making processes and research that aims to produce alternative policy proposals regarding specific policies.[3] In this sense, Suda's criticism hits the mark.

In contrast, there are academic associations in Japan with names such as the 'Society for Environmental Economics and Policy Studies' and the Japan 'Association for Environmental Law and Policy'. As the titles indicate, environmental policy studies is a central theme of the research conducted by scholars in environmental economics and law in Japan. Further evidence, if required, can be found in the titles of a pioneering text in Japan entitled *Environmental economics* (Ueta et al. 1991). This book is divided into two parts; Part 2, entitled 'Environmental Policy Studies', is wholly dedicated to a discussion of the main areas of environmental policy. Abe and Awaji (1998) is a textbook on environmental law, and their major focus is, as indicated by the title, 'Environmental Law' (*Kankyōhō*), the analysis of environmental law through the basic structure of environmental policy, administration and regulation system. Similarly, the Japan Association for Environmental Law and Policy, founded in 1997, holds a symposium during its annual conference on themes such as the environmental assessment system (1997), the legal system to deal with waste recycling (1998), natural environmental protection law (1999), and the 'PRTR (pollutant release and transfer register) Law' regarding chemicals and soil pollution (2000). These themes demonstrate the extent to which this association is highly policy-oriented.

Several Japanese environmental sociologists have become involved with local (prefectures, cities, towns and villages) governments about their environmental policies and planning. However, none have been deeply involved with the national planning authorities in the Ministry of the Environment. In contrast, environmental law scholars such as Morishima and Asano and environmental economists such as Sawa have been advisors to the Environment Ministry for many years. Environmental sociologists in Japan are, however, oriented towards practice. Article 2 of the Constitution of the Japanese Association for Environmental Sociology states that members shall 'contribute to the resolution of environmental problems'. And yet it remains unclear what contribution environmental sociology has made, or indeed can make to resolving environmental problems. Nor has this issue been fully debated, either within or outside of the Association for Environmental Sociology.

Even in the USA, where sociological studies are generally highly policy-oriented, environmental sociology also has a less effective policy-orientation. The first genuine environmental sociology textbook published in English was by Humphrey and Buttel (1982). It mentions the key features of environmental policy in the concluding chapter, but does not discuss these features in any detail. A revised edition of this work was published in 2001 (Humphrey, Lewis and Buttel 2001) does not even mention environmental policy. In contrast, both editions devote an entire chapter to environmental movements.

This policy-less orientation is not unique to environmental sociology, but is shared by sociology in general, especially in Japan. Japanese sociology has never been policy-oriented except in the areas of family and welfare. In general terms, sociology has addressed some limited aspects of public policy through the theories of social planning and social welfare, but empirical research has for the most part been limited to social welfare planning, with the notable exception of Takegawa's work on local planning (Takegawa 1992).

There was some hope that environmental sociology might contribute to the debates on the local agenda and local environmental planning based on the Agenda 21 adopted by the Earth Summit of 1992, and by developing models for assessing environmental policies and their impact. However to date, there has been very little work in this area by Japanese sociologists, with the notable exception of Seki's (2001) pioneering work.

The policy-less orientation of mainstream sociology

Japanese sociology's 'failure' to develop a policy orientation is a product of history and structure that needs to be examined from a variety of angles.

1. First is the principled and holistic nature of Japanese sociology. Popper (1957), a philosopher of science, criticized Mannheim, arguing that sociology should be 'a science' engaged in 'piecemeal social engineering' based on 'falsifiable hypotheses' rather than a 'philosophy of society' or a 'philosophy of social reform'. But Japanese sociology did not develop along the lines suggested by Popper, partly because of the strong influence of Marxism which tends towards a holistic understanding of society rather than building and elaborating explanatory models by breaking down variables.

2. To date, Japanese sociological theorists have demonstrated strong tendencies to debate earlier theories and to introduce the latest theoretical developments from the West.[4] Efforts to face and tackle the concrete social problems and policy issues confronting Japanese society and to develop theories based on primary research remain very weak.

3. Japanese sociology has emphasized empirical research—the collection of qualitative and quantitative data—but very little effort has been made to use this data to make policy proposals.

4. Although legal scholars and public administration studies scholars have a detailed understanding of Japan's various systems and laws, the major focus of empirical sociologists has been the everyday life perspectives of ordinary people—there been little interest in legal systems or policy.

5. Kōsaka has noted that the 'spell of Max Weber' and the 'spell of Karl Marx' have had very strong influences in Japanese sociology and may, at least partially, account for its policy-less orientation (Kōsaka 2000). Weber's stoic view of scholarship emphasized the importance of the 'value free' or dispassionate observer while Marxist theory is oriented towards wholesale reform of the social system. Thus both either implicitly or explicitly discourage cooperating with government policy-makers, especially in a situation where a single dominant political party has retained power for over 50 years (except for a few years in the mid 1990s).

6. In Japanese universities, sociology is often located in the faculty of arts and letters. Students from law and economics faculties (i.e. the excellent students who are most likely to become policy-makers in the national government) tend to learn sociology as an introductory liberal arts subject rather than a specialist discipline. Furthermore, within the arts faculty, sociology has

typically been located in the philosophy department, which tends to reinforce the perception that it is an abstract and speculative field of criticism.

7. Political science in Japan once shared the characteristics outlined in points (1) and (2) above. In the late 1970s, however, it rapidly changed into a policy-oriented discipline, abandoning its earlier focus on debating academic theories (Ōtake 1994). This change is generally considered to have been affected by academics such as Ōtake, who completed their postgraduate degrees in the USA. In sociology, however, academics generally tend to complete both their MA and PhD in Japanese universities, and thus the policy-orientation of mainstream American sociology has had little impact in Japan.

8. Sociological policy research requires collecting data from and interviewing the policy-making authorities. However, although there has been some movement towards institutionalizing freedom of information policies as in many Western countries, as mentioned above, Japanese policy-making has traditionally been closed to public scrutiny. Requests to national and local governments and private businesses for access to information or interviews have therefore been answered on a case-by-case basis. Thus approaches from researchers who are not members of, for instance, official commissions or advisory panels has generally tended to be cautious and defensive.

The significance of policy research in environmental sociology

Many traditionally 'abstract' disciplines in Japanese universities have begun to develop more 'concrete' practices simply to survive as disciplines. Besides political science's shift towards a policy-orientation mentioned above, philosophy departments have begun developing and promoting 'clinical philosophy' while others have adopted fieldwork and policy orientations. The ability of all academic fields to solve problems and to make policy proposals, that is, to demonstrate their 'usefulness' or practice-orientation, has begun to be widely required against the background of the rapid decline in the numbers of eighteen year olds in Japan as a result of the low birth rate and the reorganization of universities, graduate schools and scholarship. There is a growing oversupply of university places for students and a looming crisis of funding as government subsidies are correspondingly reduced. Also Japan, along with England and many other developed countries, has entered an era in which universities, graduate schools, faculties, departments and disciplines must be audited, assessed and evaluated in accordance with the practices of the business sector. Whether fortunately or unfortunately, the most influential standard of measuring performace is the "usefulness' or practice-orientation of the unit or organization. Thus for the foreseeable future, the usefulness of sociology will continue to be questioned again and again, and to a greater extent than previously.

This, of course, applies to environmental sociology too, where the central challenge will be to increase its policy-orientation, drawing upon its foundations in sociological analysis and its accumulated knowledge. The demand for this change arises on several fronts. First, activists on the frontlines of environmental problems expect that specialist researchers will contribute to concrete problem resolution through interdisciplinary cooperation. For example, in response to environmental economists' arguments for the need to internalize the social costs of environmental protection activities, Suda (in the symposium mentioned above) points out that, 'there is almost no academic research which proposes concrete means for internalizing social costs' (Awaji et al. eds, 2001: 74). Although Suda's critique is directed at environmental economics, it applies to environmental sociology as well. Such criticisms should be recognized as an expression of a strong desire from the frontlines of the environmental movements for research that is useful in reforming laws and institutions that affect the environment. This desire for concrete research outcomes can perhaps only be satisfied through interdisciplinary cooperation between environmental sociology, environmental economics and environmental law.

Second, there are very few scholars in Japan engaged in re-searching environmental politics and environmental administration. Notable exceptions include the work of Tsujinaka in Japanese environmental politics and Kameyama (Takamura and Kameyama eds, 2002; Kameyama 2003) and Kaku et al. in international environmental politics (Kaku and Maruyama eds, 1997). Environ-mental sociology should therefore pursue further research into the decision-making processes in environmental policy and the pro-cesses for implementing environmental policies at both the national and local levels of government.

Third, as we entered the 21st century there is increasing evidence of a strong public desire for substantial social/political/policy reform. The 1990s have come to be seen as 'a lost decade' and viewed with regret as the limitations of the bureaucratic and closed policy-making system have become apparent. There is a growing demand for new concepts and perspectives, and attempts to create a new public space are increasingly numerous. Authorities in charge of policy are gradually beginning to accept policy proposals from an environmental sociological point of view.

Under these circumstances, then, how can environmental sociology be of use? An examination of the characteristics of environmental sociology and sociological research into environmental problems will help to answer this question.

The characteristics of environmental sociology

In its short history, sociology has developed a vast array of research methods and theories, as well as opinions about which are 'best' or 'most sociological'. However, the important characteristics of the sociological approach include (1) understanding social phenomena as the effects of human actions, and therefore as dynamic and changeable according to people's decisions and choices, and (2) emphasizing the importance of primary research data. In contrast, economists tend to rely upon government statistics and legal scholars on legislation and court decisions: that is, both tend to analyze secondary data. Sociologists, collect and analyze primary data using a variety of research methods such as surveys and interviews. It is this type of information for which others look to sociologists. (3) Sociology also emphasizes the importance of local knowledge and the perspectives of ordinary people. In other words, 'sociological research' tends to focus on the ways in which ordinary people construct meaning and how they understand the problems that confront them. This focus includes a strong interest in local communities as a primary site of everyday life. Furthermore, (4) the sociological method is strongly oriented towards structural explanations, with the aim of developing a holistic understanding of a particular social problem.

In contrast, environmental economics and environmental law tend to focus on a single factor (they might say a 'single overriding factor') in their analyses. Environmental economists are concerned almost solely with the functions and limitations of market mechanisms. Thus, for example, they might ask how environmental values can be quantified, or how the 'external costs' of environmental protection/rectification can be internalized by an enterprise. Environmental law scholars begin with legal norms and laws as their premises, and focus on whether environmental rights have been violated or not, who is legally responsible, and how rights can be restored or legal restitution made. In both disciplines, the focus is limited and technical.

Thus sociology is unique in its recognition of diverse values and norms, and in its conviction that only by considering the full range of factors that affect local communities, cultures, individual lives, families and communities, neighborhood organizations, citizen groups and social movements can one develop an understanding of the underlying structures of a problem. The sociological analysis of the particularities and communalities of human actions, groups/organizations and communities would seem to be indispensable to formulating adequate social policy. Using the issues surrounding the bullet trains once again for a concrete example—legal scholars are only interested in the arguments that unfold in court, the possibility of proving torts, whether rights were 'violated' or not, and the effectiveness of claiming environmental rights. Economists focus only on the 'internalization of social costs', 'development effects' such as increasing enterprises, factories and populations, and 'cost-benefit analyses' of various policies that attribute costs to the source of the problem or to the surrounding area. In contrast, sociological research into bullet train related problems has been far more comprehensive, investigating four broad areas: (1) damage incurred by human victims, including both the direct effects of noise and vibration, and the less direct effects on human lives such as the multitude effects of relocation on families and communities, (2) the structural/institutional mechanisms behind the damage, such as the bureaucratic 'history of pretending to notice no wrongdoings' and the slavish devotion to precedent, (3) social movements, for example, the impact of residents' movements and legal cases, (4) government policies, including comparing perceptions of the social impact of policies, social benefits, development effects, and a critical examination of the concept of 'public works' (Funabashi et al. 1988).

The very nature of environmental problems demands inter-disciplinary research responses. The failure of academic disciplines to deal with this area has called into question the effectiveness of various existing disciplines and the ideals of existing academic research. To date, Japanese environmental sociologists have made only limited contributions to the problems of waste, energy, the global environment, and especially global warming. As discussed above, though, environmental sociology can be uniquely placed to make a significant contribution here. As the possibilities for the environmental movements to contribute to policy-making continues to open up, so too will the opportunities for environmental sociology.

The three dimensions of environmental policy

Theories of decision-making, technical methods, and values

Sociological theories of social planning and public policy identify five essential elements of social planning: the planning subject (who?), purpose (what?), object (why?), means (how?) and evaluation (did it work? was it effective?) (Kaneko 1982). Broadly speaking, social planning study consists of three areas: (1) planning decision-making, (2) planning technical methods, and (3) planning value. Environmental policy research can be similarly divided into three dimensions: (1) the decision-making process, (2) the technical methods, and (3) values.

1. In researching environmental decision-making processes, the key issues are identifying the aims, and documenting the drafting, implementation, and evaluation of plans and policies. These should be arranged in chronological order from proposal to implementation according to the processes and paths used, including the types of relationships (e.g., conflict, corporate or other) between various actors. Research in this area has included identifying the different categories of subjects (actors), the procedures involved in decision-making, whether and how consensus is achieved, environmental impact assessment, the process by which environmental problems are identified and disputed, power and participation, the role of the environmental movements (including NGOs and NPOs), democracy, the public sphere, and the responsibility for planning and policies.

2. Researchers of environmental policy technical methods primarily evaluate the various means proposed for achieving a given environmental end, including an assessment of the spillover effects and social influence. This includes analysis of the proposed methods for achieving environmental aims and policies, evaluating the appropriateness and effectiveness of proposed policies for realizing given aims, and the various resources that are available for achieving these ends. Research in this area has included examining the laws and systems governing environmental relations, environmental taxes and 'pollution vouchers', the policy mix approach, methods of internalizing social costs, CVM (Contingent Valuation

Methods), WTP (Willing To Pay), whether burdens should be shouldered by those who can pay or those who directly benefit, green purchasing, environmental auditing and accounting, life-cycle assessment, and feasibility studies. These are the areas in which environmental economics and law have primarily focused in discussing environmental policy.

3. Researching the values underlying environmental policy requires meta-analysis. Its primary objective is to explicate and evaluate the aims and postulates of environmental policy, and compare these to dominant social values, environmental ethics, and various environmental rights. Of course, comparing policy to dominant social values requires determining precisely what those social values are. Researchers in this area have addressed the values underlying environmental justice and social justice, environmental rights, nature's rights, the public sphere, policy transparency, social accountability (the responsibility to explain), amenities, PPP (Polluter Pays Principle), EPR (extended producer responsibility), and the precautionary principle. Normative analyses of ideal values and environmental policy are common. The new discipline of environmental ethics is primarily concerned with arguments in these areas (Shrader-Frechette 1981; Katö ed. 1998). Environmental law specialists have also made many contributions here, beginning with Awaji (1980). In environmental economics, Miyamoto (1989) has emphasized this perspective.

Japanese environmental sociologists have contributed to all three of these dimensions. The work of Iijima (1984/1993; 2000) and Funabashi (1990; 2000) fit into the first category (decision-making). Research conducted by Umino and others (Nakano et al. 1996) focus on factors that promote or prevent behavior to reduce garbage, and is thus concerned with the second dimension (technical aspects). Many authors have contributed to the third category (values), including Toda (1994), Ikeda (1995), Torigoe, Kada and others (Torigoe and Kada eds, 1984; Torigoe ed. 1989). It seems then that environmental sociology has been a strong contributor to the third (values) dimension, but less so on the other two. Hopefully in the future it will contribute effectively to all three dimensions. Doing so, however, will require joint research projects with environmental economists, environmental law specialists and environmental ethics specialists.

When evaluating discussions about environmental policy, it is quite useful to recognize which of these three dimensions is 'setting the agenda'. To summarize, in any discussion of environmental policy, the following points should be examined. (1) Why was that policy drafted? What environmental problems and policy issues lie in the background, and what were the decision-making processes and social interests that led to the policy? (2) To what extent are the means effective in realizing the ends? What sort of spillover effects and social impacts can be expected? (3) What are the policy ends? What are the values underlying these ends? What are the assumed rights of the stake-holders? Only when all three dimensions are clearly articulated is a comprehensive and synthetic discussion of environmental policy possible.

The principle of extended producer responsibility

As discussed, then, there are three dimensions of environmental policy to be considered, and many of the current issues fall into one of these dimensions. Here I will examine the concept of extended producer responsibility (EPR), which cuts across all three. To date, efforts to reduce the amount of waste produced and to recycle where possible have focused on distributors or consumers, waste collectors and disposers, and local and national governments. As its name suggests, EPR attributes responsibility for waste reduction and recycling directly to the producer. EPR refers to attempts to motivate the producer to reduce the amount of waste produced, to sell products that can be recycled, and to establish recycling networks. Its principal aim is to make producers and manufacturers bear the costs of waste, on the principle that the most effective place to reduce waste is to impose restrictions at the top of the production-distribution-consumption-waste chain. Since the producer profits from this chain, imposing these downstream costs on the upstream producer can be seen as an application of the 'beneficiary pays' principle. In Japan, from April 2000, the Containers and Packaging Recycling Law which was implemented for all containers and packaging mentioned only the responsibility of the manufacturer to recycle. However, beginning with Germany's Closed Substance Cycle and Waste Management Act of 1994, similar systems are widely practiced in the EU, and are being promoted by the OECD.

Research into the basic principles, ideals, and values of EPR is concerned with the value dimension of policy. Questions also arise on the technical method dimension: how to design a workable system in accordance with the EPR principle. And an analysis of decision-making processes is required to show why this principle was put into practice in Germany and other EU nations but not in Japan (yet). Germany has a population of 82,070,000, of which 550,000 are members of Greenpeace (or 0.7% of the total population) and 360,000 are patrons of the WWF (0.4%).[5] There are also 250,000 members of the national environmental body BUND (Bund fur Umwelt und Naturschutz Deutschland, 0.3%). The Netherlands has a population of 15,600,000, of which 600,000 are members of Greenpeace (3.8% of the total population) and 830,000 are members of the WWF (5.3%). The Netherlands' percentages are 5 to 12 times higher than those of Germany. In contrast, Japan, with a population of 120,000,000 had a mere 5,000 and 40,000 members respectively in these two organizations (all figures as of 1997). The social influence of environmental NGOs is thus very different in EU nations and Japan, and the distance between government and NGOs is much larger in Japan than the EU countries.[6] That is, EU governments are far more responsive to NGO policy proposals than in Japan.

The role of the environmental movements in environmental policy

To discuss the 'dynamism of the environmental movements and environmental policy' is to emphasize the role of the environmental movements as influential actors in three dimensions; (1) the policy-making process, (2) evaluating policy means, and (3) proposing policy values.

Returning to the bullet trains example (see chapter 7), the protest movement against the bullet train pollution in Nagoya publicized the extent of the problem through a residents' movement and legal action, and prompted the authorities to formulate a response; the movement influenced the decision-making process. However, the initial response of the Japan National Railways was very limited; what the movement most desired was that the bullet trains slow down when passing through a 7km length of residential areas (evaluation of policy means). This demand was supported by the JNR labor unions, and for nine years from 1974 approximately two-thirds of the bullet trains that passed through the contested areas did in fact slow down. The protests against bullet train pollution directly opposed the JNR's expressed value of high-speed mass transportation with an alternative value, the residents' right to 'peace and quiet' (a policy value proposal). However, since the protest movement lost two court cases, the trains have returned to full-speed operations.

7   Anti-pollution lawsuits in the public sphere

Introduction

The Nagoya bullet train and Osaka Airport lawsuits were the two largest anti-pollution lawsuits of the 1970s, and greatly influenced the direction of later trials. As described in chapter 3, 'high-speed transportation pollution' was typical of the environmental problems that surfaced during the 1970s. This chapter will consider the following issues.

What has been the social significance of particular anti-pollution lawsuits? What are the general social functions of anti-pollution lawsuits? Do they contribute to ameliorating damages and settling disputes, and if so, by what mechanisms?

What was the social background to the tendency that emerged in the 1980s for judicial passivity in decisions concerning environmental lawsuits? What are the benefits and limitations of the 'lawsuit strategy' for local residents' movements? What problems does this strategy pose for the residents' movements? How do the movements themselves change once the dispute has moved to court?

In the mid 1980s, when the first version of this chapter was published, there had been little sociological discussion of the legal or political problems of anti-pollution lawsuits. Since then, these discussions have not increased to the extent that I had then expected.[1]

Certainly there has been vigorous discussion in the mass media and specialist legal journals about major lawsuits. Each ruling is widely reported, including comments by the defense and plaintiffs' attorneys and 'expert opinions' about the rulings. Editorials and expert opinions significantly influence public opinion. Yet this type of commentary is almost invariably limited to legalistic debates about the merits of a case and the appropriateness of the ruling.

To understand the full social significance and function of these cases, however, a review of the rulings alone is not sufficient. In the Osaka Airport and Nagoya bullet train cases, for example, the act of filing the lawsuits had major social impacts—arousing public opinion and forcing policy makers and responsible parties to implement various counter measures. The rulings also had more direct benefits for the particular plaintiffs involved, but even when they were successful, this was merely a part of the overall result achieved by pursuing the lawsuits strategy. Since the 1980s, an increasing number of cases are resolved through out-of-court settlements. It is therefore necessary to consider the entire process of each case to determine how the lawsuit affected the plaintiffs and defendants, the authorities concerned and public opinion.

Of course, each case is unique and the judicial ruling in a specific lawsuit cannot be simply generalized without qualification. However, through a comparative study of some of the major anti-pollution lawsuits it is possible to delineate between the particular factors and more general factors. From there, it is possible to discuss the matter in general terms.

In this chapter, drawing on primary data collected during a field survey I conducted from 1981 to 1985 with Funabashi and others (Funabashi et al. 1985), and the available secondary materials about the Nagoya bullet train and Osaka Airport cases,[2] I will present a generalized outline of the legal proceedings and rulings and discuss their broad implications from a sociological perspective.

'Contemporary lawsuits'

The Osaka International Airport pollution lawsuit (filed in 1969— hereafter the Osaka Airport case) and the bullet train pollution lawsuit in Nagoya (filed in 1974—hereafter the bullet train case) are early examples of a new type of lawsuit in Japan variously called 'contemporary lawsuits' or 'actions for policy formation'. These lawsuits were unique because first, the issue in dispute concerned the basic values of public policy (Rokumoto 1991). Second, they were the collective actions of large numbers of plaintiffs. Third, legal action was consciously adopted as a strategy by the plaintiffs, both to resolve individual disputes and to achieve the larger objectives of the social movements. In both cases, such a strategy was necessary due to important structural features of the Japanese bureaucracy and public works administration: there were no formal channels available for affected citizens to challenge the operating procedures of public works bodies and other government agencies or to object to the construction of new facilities, which left the judiciary as the only avenue available for airing and resolving citizens' and residents' grievances against the decisions and operations of public authorities.

In this sense the judicial system served as a formalized 'public sphere' (as defined in chapter 11) where social problems could be aired and debated. The court thus provides a venue in which ordinary citizens can make claims about the nature and extent of pollution damage or prospective damages on an equal footing as the defendant, and thereby appeal to the broader society for justice, compensation and resolution of the environmental problem.[3]

Judicial trial as a 'resource mobilizing process'

A typical proposition in the resource mobilization perspective is that 'protest is correctly conceived as a strategy utilized by relatively powerless groups in order to increase their bargaining ability' (Lipsky, 1968: 1157). Lipsky defined 'relatively powerless groups' as 'those groups which, relatively speaking, are lacking in conventional political resources' (Lipsky, 1968: 1144). Following this proposition, we can say that filing a lawsuit is correctly conceived as a strategy utilized by relatively powerless groups in order to increase the possibility of mobilizing social resources and their bargaining ability. For social movements, litigation can be a tool for mobilizing resources. Thus, residents' movements with no other routes for negotiating their grievances are frequently inclined towards strategic litigation.

However the courtroom is a very peculiar public sphere with many limitations. For example, the judge presides as a transcendent authority and arbiter over the disputants, arguments about the disputed 'rights' are conducted according to procedural laws in a very technical and formal manner, and the judge is restrained by precedents (i.e., the judge is not the ultimate authority).

Road closed to injunctions

On 12 April 1985, the Nagoya High Court handed down its decision on an appeal in the bullet train lawsuit, upholding the original trial judge's decision (September 1980) to reject the plaintiffs' application for an injunction against the Japan National Railway (JNR) to reduce bullet train noise and vibrations. Following the Supreme Court's decision (December 1981) to dismiss the Osaka Airport lawsuit, the possibility of achieving an injunction against a large public works project or public facility seemed remote. In fact, there was no precedent for an injunction order against a large-scale public works project or facility until the January 2000 ruling in favor of

Figure 7.1   The most severely damaged area of Tökaidö bullet train pollution in Nagoya City, called '7km stretch of densely populated area'

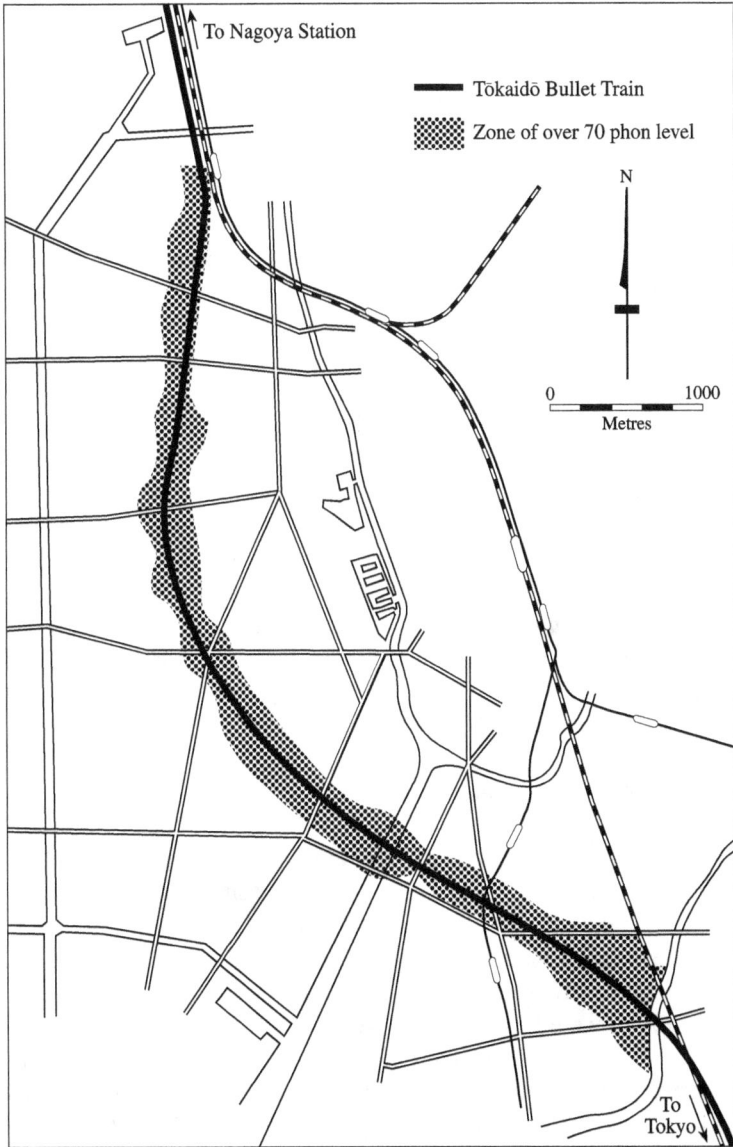

To Nagoya Station

Tōkaidō Bullet Train

Zone of over 70 phon level

N

0   1000
Metres

To Nagoya Station

To Tokyo

Nore: Adapted from Funabashi et al. (1985: xii)

the plaintiffs in the District Court trial of the Amagasaki pollution lawsuit and the November 2000 ruling in the initial trial of the lawsuit over air pollution in southern Nagoya, a complex air pollution problem produced by a combination of car exhaust fumes and factory emissions. In both cases partial injunctions were ordered against the discharge of air pollutants. (Both cases were settled out-of-court during the appeal process to the High Court.)

The Supreme Court decision in the Osaka Airport case overruled the Osaka High Court decision (November 1975) to fully accept the plaintiffs' claim, stating that where the State was the founder and proprietor of a facility, there was no legal basis to seek an injunction against the facility by civil litigation. In the bullet train lawsuit, neither the original nor appeal court decisions recognized a causal relationship between the pollution and the physical damages claimed. The 'public' nature of the bullet train was cited and the damages were determined to be within the bearable limits defined in the injunction. The injunction was therefore rejected.

Both the Osaka Airport and bullet train lawsuits were complaints against large-scale and serious pollution and had long histories of damages. They were the first Japanese lawsuits to be filed by large numbers of plaintiffs. When they were lodged, litigation seemed to offer the residents the best opportunity for achieving an injunction, thus their failures were highly significant. In both cases the judiciary avoided ordering an injunction that would effectively impose regulatory standards on public works, narrowly defining its own role in resolving pollution disputes or awarding compensation for damages suffered.

The plaintiffs' demands

In both cases, the plaintiffs' demands were not very difficult to achieve. In fact, they were quite minor when compared to the severity of the damages already and continuing to be incurred.

In the Osaka Airport case, the plaintiffs sought an injunction banning take-offs and landings between 9 pm and 7 am. Operations after 10 pm were banned by the Ministry of Transportation in May 1972, so by the time the High Court found in favor of the plaintiffs in 1975 the matter in dispute was merely one hour of night operations. Following the High Court decision, administrative regulations were adopted to impose a 9 pm curfew on flight operations. Six years later, when the Supreme Court overturned the High Court's ruling, these regulations remained and are still in place today. In fact, a commitment to continue the 9 pm curfew was a key factor in the settlements of the 4th and 5th lawsuits in March 1984.[4]

In the bullet train case, the plaintiffs sought an injunction to limit the trains' operating speed to 110km/hour to decrease the noise and vibration levels through a 7km stretch of densely populated area (70km/hour had been demanded in the District Court trial). But from February 1974, just before the lawsuit was filed, until May 1983, almost for nine years, the National Railway Motive Power Union (Dörö) had instructed its members to operate at around 110km/hour in this stretch. This did not significantly affect the train timetable due to built-in allowances in the schedule. Thus although the JNR managed to pressure the unions to return to full-speed operations, there was once again very little at stake in conceding to the plaintiffs' demands.

In both cases, the plaintiffs' demands had already been substantially met for many years without complication. Hence, if the decisions had recognized the demands, the affects for the defendant (the 'public' in effect) would have been minimal compared to the benefits of reducing damage to the affected areas and the benefits for both parties in finally resolving the disputes. The court could have legally sanctioned these outcomes, but in both cases chose 'judicial self-suppression' over the interests of the plaintiffs, adopting a 'passive position on judicial remedies'.

However, in these first years of the 21st century the judiciary has begun to play a far more active role in anti-pollution lawsuits after the 'long winter' of about twenty years (since the early 1980s) without any prominent innovative rulings. For example, in a lawsuit (filed June 1994) demanding that the Ministry's approval for a project to elevate a section of the Odakyu Line be overturned, on 3 October 2001 the Tokyo District Court noted the Ministry's failure to properly assess the proposal, including their failure to consider measures to prevent noise pollution. It found in favor of the resident plaintiffs and handed down a landmark ruling nullifying the approval by the head of Kanto Regional Development Bureau of the Ministry of Land, Infrastructure and Transport. This nullification by administrative lawsuit of a large public works project that was already built was a landmark case in Japan. On 18 December 2002, in a case (filed in March 2001) in which local residents sued a construction company, the Tokyo District Court ruled in favor of the plaintiffs and handed down the first Japanese decision to order, on the grounds of protecting the cityscape, the physical removal of the top 20m of a high-rise unit block (in Kunitachi City, Metropolis of Tokyo). In an appeal against the decision in an administrative lawsuit, on 28 January 2003 the Nagoya High Court Kanazawa Branch ruled to nullify the planning approval for the installation of the fast breeder reactor, Monju (which was shut-down after a sodium leakage accident in 1995), on the grounds that errors had been made in its safety assessment. This was the first Japanese judgment in favor of the resident plaintiffs in a nuclear facility-related lawsuit. Only time will tell whether these judgments are exceptional and provisional or herald a new century with significantly different judicial rulings on anti-pollution lawsuits.

The significance and limitation of anti-pollution lawsuits

The Osaka Airport lawsuit came to an end in March 1984, when the 4th and 5th lawsuits were resolved with settlements. First filed in December 1969, the five lawsuits had extended over a period of 15 years (see note 4).

The JNR settled the bullet train litigation with the plaintiffs in April 1986 just one year before it was divided and privatized. The principal conditions of the settlement were (1) that the JNR would endeavor to reduce noise levels to 75dB by the end of March 1991, and (2) would pay 480 million yen in compensation. This financial settlement was very close to the 530 million yen awarded in the original trial decision; the second trial had reduced it to about 300 million yen. Twelve years had passed since the first suit was filed in March 1974.

This case was an early instance of what soon became a marked tendency for anti-pollution lawsuits to be quite protracted. From the beginning of the 1980s there was also a sharp increase in the number of rulings that imposed harsh restrictions on the plaintiffs' eligibility for awarded compensation, for example, where plaintiffs whose exposure to pollution was below the 'tolerance limit' were ineligible. This type of limited compensation was awarded in cases such as the initial trial in the Yokota military base pollution lawsuit (July 1981), the initial trial in the Atsugi military base noise pollution lawsuit (August 1982) or the Nagoya High Court ruling on the appeal in the bullet train lawsuit, where the compensation awarded by the District Court was substantially reduced.

Thus by the late 1980s anti-pollution social movements had to reassess the efficacy and limitations of litigation as a means to achieving their objectives.

Social significance of the Nagoya bullet train lawsuit

Collective action lawsuits demanding injunctions

In March 1974, at the height of the Nagoya bullet train anti-pollution struggle, 575 local residents (the plaintiffs) filed suit against the Japan National Railway (JNR) (the defendant). The plaintiffs' demands included reducing the noise and vibrations caused by the bullet train in their residential area to 65dB and 0.5mm/sec respectively during the day and 55dB and 0.3mm/sec respectively at night. Historically, the case for civil litigation followed in the wake of the four major anti-pollution lawsuits (filed in 1967–69, judgments handed down in 1971–73, as mentioned in chapter 1), and with the Osaka Airport lawsuit (1969–84) established new legal precedents for administrative litigation. The long drawn out legal process of the Nagoya bullet train lawsuit became a prime example of the tendencies and characteristics of anti-pollution lawsuits in the 1970s and 1980s, and provided a strategic model for litigation by the many residents' anti-pollution movements that would soon follow.

It thus provides a good case study for examining the potential benefits and limitations of pursuing collective objectives through legal action in the public sphere, focusing on: (1) filing suit, (2) the legal process, (3) judicial processes and judgments, (4) its social effect, and (5) the 'ripple-effect' on law-making and administrative processes.

The plaintiffs were seeking a judicial injunction against a national high-speed transport facility that was already in operation. There are five key points to note here: (i) it was a forerunner for the post-1970s pollution disputes that extended beyond the typical industrial pollution incidents and into everyday life pollutants, starting with transport pollution, to large-scale development issues such as the construction of huge industrial complexes and or nuclear power stations, food and drug contamination, neighborhood and nightclub (Karaoke) noise and the 'right to sunshine'. (ii) The defendant was the JNR—a public enterprise. With the Osaka Airport case, this was one of the first times that a residents' or citizens' movement had sought a legal judgment against a government authority to resolve or rectify pollution problems or large-scale development issues. (iii) The plaintiffs' principal demand was for regulatory measures to stop or restrict pollution generating activities, rather than *ex post facto* financial 'compensation'. In other words, they sought a direct social regulation in the form of a judicial injunction order. (iv) Personal and environmental rights were claimed to establish a legal basis for the injunction, which was not expressly defined in positive law. Following the Osaka Airport lawsuit, the case established a precedent for applications for injunctions based on personal and environmental rights. Finally, (v) the compensation sought was not only for damages already incurred, but also for the damages that would continue to be incurred until the injunction was ordered. The aim was to pressure the defendant into accelerating measures to prevent pollution through the threat of cumulative damage compensation.

Organizing collective action and outside support

In cases of transport pollution, large numbers of residents in the vicinity of the transportation facility are affected by noise etc. There were a total of 575 plaintiffs in the Nagoya bullet train lawsuit when it was filed. The first Osaka Airport lawsuit had 31 plaintiffs. The plaintiffs for the combined second and third lawsuits totaled 233, and the fourth had as many as 3,694 plaintiffs (cf. note 4). As the number of plaintiffs grew, various methods and tactics were developed to achieve smooth progress in the actions. These new methods and tactics soon became standard practices in other lawsuits.

In the four major anti-pollution lawsuits, physical damages were apparent in the form of specific diseases that were clearly caused by chemicals such as organic mercury or cadmium. In transportation pollution problems, however, health damage is often limited to stress factors, and no causative substance can be identified. It is therefore difficult to prove a causal relationship for associated health damage. Although the Public Health Department of Nagoya University Medical School, commissioned by Nagoya City, played an important role in the bullet train case by carrying out health surveys (March 1975) and analyzing the physical damages with epidemiological methods, the judicial decisions twice rejected the claims of a causal relationship. Some legal scholars criticized these judgments and voiced suspicions that the judges had intentionally rejected research results that indicated that there was, in fact, a causal relationship in order to avoid deciding in favor of an injunction. The reasoning was that if the court accepted evidence of a causal relationship once, it would henceforth have to order injunctions based on the legal precedent.

To garner public support and maintain the momentum in long term litigation with many plaintiffs, it is necessary to sustain cooperative and supportive relationships both within and outside of the court between the local residents, lawyers, scientists, national organizations of pollution victims, and sympathetic organizations such as labor unions. Strategies for organizing such relationships had been developed during the four major anti-pollution lawsuits and were further refined during the Osaka Airport and bullet train cases, soon becoming well established.

One significant development in the bullet train case was the supporting actions from two major labor unions that represented employees of the defendant. Both the National Railway Motive Power Union (Dörö) and the National Railway Worker's Union (Kokurö) attempted to ameliorate the damages by imposing restrictions on operating speeds along specified sections of the track, the former continually for more than 9 years and the latter repeatedly for limited periods. Members of both unions testified as expert witnesses on the particularities of the densely populated Nagoya area that necessitated reduced speeds, the ease of reducing speeds and its marginal impact in operational efficiency. Representatives of both unions also attended plaintiffs' meetings and the negotiations between the JNR and the plaintiffs, and engaged in various supporting activities. This type of long-term proactive support by the labor unions of a defendant company is quite rare in the history of Japanese pollution issues.

The social significance of the bullet train ruling

Except in the High Court decision (total injunction) and the District Court decision (partial injunction) for the Osaka Airport lawsuit, the judiciary has not ordered an injunction against any large public facility until quite recently. Instead, the judicial relief system was established in the form of damage compensation. In the four major anti-pollution lawsuits, the legal responsibilities of the private (corporate) defendants were recognized. In the Osaka Airport case the courts recognized the State's legal responsibility—as the airport founder and proprietor—for the airport's deficiencies. The bullet train case was the first time that the courts acknowledged the legal responsibility of a public enterprise for damages resulting from its operations. This affirmed that even public facilities with high social utility could be held legally responsible. The two latter cases also set precedents whereby damages to 'everyday life', such as disturbances to conversation and sleep by excessive noise and vibration, were added to the list of pollution related complaints for which one might seek legal restitution or compensation.

The social influence of particular lawsuits

The four major pollution problems and their court proceedings aroused public concern about pollution issues and raised questions about the government's economic growth priorities. They also opened a path to judicially awarded fiscal compensation for the victims of environmental pollution and contamination that stimulated subsequent lawsuits. The bullet train and Osaka Airport cases made the public aware that the comfort and convenience of high speed transportation came at the cost of loud noise and vibrations, and that this cost was born by the residents of the areas around the facilities who were forced to endure it because of inadequate legislation and negligent administrative processes. This new awareness led to widespread reflections on an increasingly high-speed lifestyle and society. The cases also made many people aware for the first time of the limitations of technology for resolving social issues as it became increasingly clear that in spite of their demonstrated ability to develop and operate world-leading transportation technology, the JNR could not 'fix' the accompanying noise and vibration problems. And they demonstrated that the physical and psychological damage of high-speed transportation pollution are social issues, inducing the formulation of social-values critical of high-speed society.

As the public became increasingly aware of its right to be free of the effects of pollution, there was a growing social expectation that the judicial system would protect these rights and compensate individuals whose rights had been violated. At the same time plaintiffs' lawyers were refining their techniques for pursuing large-scale collective action lawsuits, and the lawsuits strategy became the principal method through which residents' movements' fought for an end to their suffering (or the prevention of foreseeable suffering). Similar cases followed in ever-increasing numbers, including lawsuits involving airport pollution and military base pollution,[5] the administrative litigation by Urawa City residents demanding an injunction against the construction of the Tōhoku and Jōetsu bullet train lines, and civil litigation by residents in Kita Ward, Metropolis of Tokyo, demanding an injunction for the construction of Tōhoku and Jōetsu bullet train lines.[6]

Influencing legislative and administrative processes

The four major anti-pollution lawsuits were a turning point in anti-pollution legislation. The Basic Law on Pollution Control (enacted in 1967) was amended in 1970 at 'The Pollution Diet' (the 64th Extraordinary Diet Session). Behind the growing pollution problems and increased litigation was deficient legislation and a dysfunctional administration, with its closed bureaucracy and long overdue administrative grievance resolution mechanisms. Publicly airing these grievances in court aroused widespread interest and public pressure, prompting legislative changes and the formulation of new administrative institutions. For example, in 1971 the Environmental Agency was established. The Osaka Airport lawsuit (December 1969) prompted legislative and administrative measures such as: the Aircraft Noise Environmental Quality Standard (December 1973); amendments to the Aircraft Noise Pollution Control Law (enacted August 1967, amended March 1974); the creation of the Organization for Environmental Improvement around the Osaka International Airport within the Osaka Prefectural Head Office (April 1974) to dispense compensation for relocation and to implement corrective measures such as soundproofing; and regulations to limit the take-off and landing times and the number of flights at the Osaka Airport. Filing the Nagoya bullet train lawsuit (March 1974) also prompted a number of administrative measures including the Bullet Train Noise Environmental Quality Standard (July 1975), Bullet Train Vibration Emergency Index recommendation (March 1976), and Cabinet agreement on the Bullet Train Noise Control Outline (March 1976) which became the basic rules for pollution prevention measures such as soundproofing and compensation for relocation.

Judicial processes and their influence on society

General social functions of the lawsuit

From these specific cases we can identify the general social significance and social functions of anti-pollution lawsuits. First, these lawsuits function to frame the pollution problem as a social

issue. They inform the broader society of the existence and extent
of pollution damage. They also identify the (alleged) responsible
parties, the shortcomings of government structures, and the urgency
of redress or countermeasures.

Second, these lawsuits function to (re-)form institutions and
practices. This has been called the 'policymaking function of the
judicial trial' (Tanaka 1979:118). Breaking this function down by
the levels of institutions and standards, we see (i) the formation/
reformation of institutions as administrative and judicial measures,
(ii) creating legal precedents, (iii) extending and reinforcing
rights, as in the growing recognition of 'environmental rights', and
(iv) transforming general public attitudes towards anti-pollution
measures and damage relief.

Third, these cases function to provide damage relief. The judgment has not only an *ex post facto* damage relief effect for the
plaintiffs, but as we have seen, new institutions sometimes emerge,
or the ruling has a ripple effect that benefits victims of similar
pollution in other places. The ripple effect may produce stricter
regulations being imposed upon planned facilities and those already
under construction. It may stimulate further anti-pollution lawsuits
in other districts as the anti-pollution movements in those areas seek
progress in environmental protection measures. A good example
is the advances in the pollution measures of the new Tōhoku and
Jōetsu bullet trains, where in the wake of the Nagoya lawsuit and
under pressure from the anti-construction movement along the new
lines, innovative anti-noise measures were employed according to the
particular circumstances along each particular affected section of the
line. However, even on these lines, the operator attempted to increase
the operating speeds even before the environmental standard of 70dB
in residential areas had been achieved along the entire line, making
the standard (but not the measures employed to meet them) rather
meaningless (Funabashi et al. 1988: 75–6).

What institutional characteristics and socio-political mechanisms
of the judicial process, then, provide the bases for these social
functions? These will be examined separately as (i) the processes
until the lawsuit is filed, and (ii) the judgment process.

Until the lawsuit is filed

There is a common pattern in the processes that lead residents'
movements to file a lawsuit. The local residents' movement typically

begins to seek relief from pollution damage by lobbying and protesting. As they tire of these direct and indirect activities they often begin to feel frustrated by the ineffectiveness of their actions, develop distrust of both the ruling and opposition parties, and feel like they are 'up against the wall'. Only then do they consider taking the case to court.

They consult with lawyers to discuss the possibilities and merits of court action. What might be achieved, what are the chances of a victory (especially against the State or its agencies), what will the cost be, and how long will it take? These are only some of the questions. Scientific and health experts must also be consulted—both to interview/examine the residents to assess the nature and extent of the damage, and to respond to their doubts and worries.

Of course, often the lawyers themselves lead the residents' group to court. The anti-pollution committees of the Japan Federation of Bar Associations as well as the regional and district bar associations are quite proactive in such matters. In both the Osaka Airport and Nagoya bullet train cases, the respective bar association's own investigations into the pollution damage led the lawyers to initiate meetings with suffering local residents. The lawyers also took the initiative in the various lawsuits against the Tōhoku and Jōetsu bullet trains. Their 'preliminary' investigations in these instances were effectively preparations for the lawsuits.

Various roles of the lawyers

Once the lawsuit strategy has been decided upon, the social resources of the plaintiffs' counsel can be systematically utilized until the confirmation of the final judgment. The roles of a plaintiffs' counsel are various and comprehensive both in and out of the courtroom, as described below.[7] In sociological terms, during a judicial trial the plaintiffs' counsel could be construed as one of the competing epicenters in a resource mobilizing process.

In large-scale anti-pollution lawsuits about complex issues requiring large numbers of witnesses and even larger numbers of evidentiary documents, many lawyers are required to prepare for and fight the case. Organizing the lawyers and assigning clearly defined roles is therefore the first necessity. The particular lawyers who undertook the damage investigation often play the key roles in organizing the plaintiffs' counsel. The counsel listens to the residents' complaints and desires, researches legal precedents, related lawsuits and academic remarks, determines the specific details of the application for an injunction, its basis in law and the amount of monetary compensation to be demanded, then prepares the legal documents and a strategy for fighting the case.

As the final plans and documents materialize, the counsel works with the movements' leaders to call for formal plaintiffs, and a plaintiffs' group is organized. Where the plaintiffs' group is born from an existing residents' movement organization, plaintiffs must still be sought to assess the individual residents' cases, and to consider the extent of the damage, determine the total numbers of plaintiffs and each individual's intention to continue with the lawsuit.

Once the plaintiffs are confirmed, the complaint is submitted and the trial begins. The plaintiffs' counsel must perform numerous tasks/roles in the courtroom. (a) The counsel must demonstrate the justification for the plaintiffs' demands for relief based in positive law. Where the justification is not explicitly enshrined in law, they must use broader legal principles, such as personal rights and survival rights, to argue their case. The plaintiffs' earnest wish to live in 'peace and quiet' and their anger at the 'insincerity' of the defendant who (allegedly) left the pollution unchecked are enhanced to support the demands of rights. This is well illustrated by the Osaka Bar Association's proposal of environmental rights at an annual conference of the Japan Federation of Bar Associations while the Osaka Airport trials were underway.[8] Seeking new legal arguments and rights claims in turn could develop a stronger legal theory and a real expansion of citizens' rights.[9]

(b) The counsel must prove the extent of the damages, the defendant's liability, the facilities' deficiency and the urgency of redress. To achieve this, information is collected from the plaintiffs and others concerned, witnesses are selected, and doctors, academics and other specialists are called as expert witnesses. Documents such as damage investigation reports, reference materials and newspaper articles are also gathered, analyzed and submitted as written evidence and so on.[10] (c) At the same time the defendant's claims are challenged and weaknesses in their arguments are drawn out through the cross-examination of their witness. (d) Throughout the proceedings the counsel must assess the defendant's preparatory documents, the evidence they present, and negotiate a trial schedule with the defendant's lawyers. In the process they attempt to deduce the defendant's position, their attitude towards the suit and resolving the dispute (which may include directly asking the defense counsel about their intentions, such as, say, to resolve the case through mediation), and their court strategy. A strategy for court is then decided. The plaintiffs' counsel also has a variety of roles outside the courtroom. (e) They must discuss the strategies and objectives for the lawsuit and the movement with the plaintiffs' group leaders. (f) They must keep the general plaintiffs informed of the progress of the case, including the court proceedings and the prospects of winning—through meetings, study seminars and newsletters etc. (g) They should vigorously support the movement leaders in maintaining the unity of the plaintiffs' group and its organization. (h) Where there are direct confrontations between the residents and the defendant's representatives, the counsel must be present and restrain the defendants. (i) They must respond to newspaper and television journalists and explain the nature of the damage and details of the lawsuit in legal journal articles, highlighting the disputed points while demonstrating the defendant's liability.

The role of plaintiffs' counsel in an anti-pollution lawsuit demands a pioneering spirit and a large investment of time and energy, but it also provides a rare learning opportunity for the lawyers involved.

Ripple effect of filing actions: (1) Presentation of the social issues

As a means of presenting and debating social issues, a lawsuit has four key characteristics: (i) that both the plaintiffs and defendants argue their cases on an equal footing, (ii) the shared objective is to clarify the defendants' liability and the legal status of their actions. To achieve this, (iii) all relevant information is disclosed and shared, and (iv) these points are guaranteed by adjective law.

The biggest issues in the trial, such as the nature and extent of the damage and liability, are also the focal points of the social problem. Before the suit is filed, the defendant can refuse requests for information and documents from residents' groups and other concerned parties with little or no repercussions. But refusal to submit requested documents to a court can give judges a bad impression, and thus the court can function as a forum for sharing information, as previously suppressed information belonging to the defendant operator/ administrator is made public. Through the examination and cross examination of witnesses and evidence, the trial also serves as a public hearing. This characteristic is noted by Kimura and Kuboi: 'With the legal procedures and publicity as security, it can be said that in Japan, "a public hearing, in its true meaning, is found in the

court"' (Kimura and Kuboi 1978: 60). Before any official information disclosure systems were established in Japan, the court was the
only institutionally guaranteed place in the 'public sphere' where the
plaintiffs could argue with the defendants on an equal footing to establish the truth about the damage claims and the urgency for taking
counter measures.

The performance of these functions is considerably enhanced when
the details of the claims, the major issues, process and background
to the filing are reported by the mass media at major points during
the case, such as the petition for an injunction, the conclusion of the
hearings, and the decision. The mass media can bring the pollution
problems and the local residents' movement to the attention of a
national audience.

Ripple effect of filing actions: (2) Defensive anti-pollution measures

Being sued tends to heighten the defendant's focus on the matters
in dispute and typically arouses two somewhat contradictory but
complementary responses: (1) a defensive attitude and determination
to mobilize whatever resources are necessary to prove its innocence
in court, and (2) a proactive stance out-of-court, taking steps to
minimize the damage to the company/government agency—including
the financial and operational costs and the damage to its reputation
that can result from a court case. These positions may overlap or
diverge in complex ways, depending on the prospects of the lawsuit,
and will be key in determining the defendant's strategy for the case.

When the defendant is a private company, for whom profitability
is the ultimate concern, apprehension about the costs of a trial often
motivates efforts to achieve an early (out-of-court) settlement. In
contrast, for the State or a government entity, the 'national' dignity
and the ripple effect for other public entities/facilities are the
primary concerns. From this perspective the financial cost of the
trial is of little importance, and it is unlikely that anybody will take
an initiative towards early resolution. In other words, where the
defendant is the State or one of its agencies, their primary objective
is to be exonerated by the court. They therefore generally assume
an inflexible stance in court, stubbornly defending themselves with
complicated legal arguments and remain unconcerned about the
prospects of a long drawn out trial.

At the same time, outside the court, public sector defendants are
likely to begin cooperating with relevant bodies/agencies to develop

and/or implement environmental measures, either technical or regulatory as appropriate.

Making progress in environmental protection measures has two key benefits for the defendant. First, as a strategic measure towards legal exoneration from liability—the court's decision is based on the legal facts, which can continue to accumulate until the conclusion of the trial. Real achievements in preventative measures until then favorably influence the judges. Prolonging the trial can therefore be a useful tactic, allowing more time for such countermeasures to be implemented and become effective.

Second, effective (and sometimes even token) measures may help to avoid social approbation by demonstrating to the public that the defendant is acting in good faith and in the public interest. As the plaintiffs' argument is presented in court and reported in the media, social pressure increases for the government entity to alleviate the victims' suffering. The defendant's faults and the government's competence will be questioned by opposition party members in the Diet and discussed in the media. Positive and effective measures to ameliorate the pollution problem itself (and thereby ward-off public criticism) are therefore indispensable to a government defendant's defense strategy.

Thus a rigid position in the courtroom, including various tactics to delay and otherwise extend the trial, and the implementation of concrete measures out-of-court are common strategies for government defendants. This double-edged response was well illustrated by the National Railway's approach to the bullet train case.

What is brought about by the ruling?

A legal judgment has numerous effects, both for the plaintiff and the defendant.

Aiding the plaintiff

If the plaintiffs' demands are accepted, the judgment can have a direct benefit for the plaintiff. Even if the case continues through appeals, and the judgment is therefore not final, if a provisional execution order is awarded, the legal force of the plaintiffs' case will have been demonstrated.

If a provisional execution of compensatory damages is awarded, as in the District Court judgment in the bullet train case, the interest earned can be put towards the appeal expenses.[11] When there is a large financial award, the interest accruing can itself be substantial, reducing the individual plaintiffs' burdens and making it easier to continue the legal action.

If the injunction sought is awarded on a provisional execution order, as in the District and High Court judgments in the Osaka Airport lawsuit, the relief from further damage is effective over the entire affected area immediately after the defendant accepts the injunction order.

Sanctioning the defendant
Obviously, any judicial award for compensation or an injunction order is a legal sanction of the defendant. Depending on how this is reported by the mass-media, the legal sanction can increase the social sanction.

Ripple effect of the ruling: Pressure on the defendant to save the victims

Rising pressure for plaintiff aid
Since the four major anti-pollution lawsuits, immediately after a judicial ruling it has become quite common for the residents' *cum* plaintiffs' organization to directly negotiate with the defendant, pressing for concessions. Often this includes many demands that could not be included in the lawsuit and were suspended until the ruling. Both a ruling in the plaintiffs' favor and the judges' reasoning for ruling in the defendant's favor can recognize the defendant's liability—moral or social liability if not legal. Either judgment might therefore be useful in pressuring the defendant to make concessions. For example, in direct negotiations immediately after the Supreme Court decision in the Osaka Airport case, the Transport Minister responsible for the airport promised to continue the existing 9 pm curfew. Although the plaintiffs 'lost' the decision a practical outcome was achieved. Appealing a decision may also motivate the defendant to adopt new anti-pollution measures, for the same reasons that the original lawsuit application might, as discussed above.

Reforms to codes and institutions
The ruling becomes part of the legal code, either setting a new or affirming an existing precedent. In the reasoning for the Supreme Court decision for the Osaka Airport case, Judge Shigemitsu Dantö explained in his dissenting view that 'constructive law-making

through precedents' was the correct way to resolve such issues. If the plaintiffs' rights and their methods for proving damages are recognized in the reasoning of the decision, they will also greatly influence later cases. Efforts by the plaintiffs' counsel to establish new legal or evidentiary standards may be justified by the ruling, and this justification is reinforced as the case becomes a precedent. (b) The decision may also prompt new legislation or administrative institutions. (c) The extent and nature of the decision's damage relief component heightens the rights awareness of similar victims and the general public, and anti-pollution movements and lawsuits become ever more active and constructive in pursuing and achieving pollution countermeasures.

Sources of the passive position on judicial remedy

As mentioned, the judiciary maintained a passive position towards creating new precedents in anti-pollution lawsuit decisions for a long time. This stance has been called 'judicial self-limitation' and 'judicial negativism' in relation to cases about national security issues or imbalances in the electoral quota; but why was it also adopted towards pollution issues? We must briefly examine the social and political background to explain this phenomenon (cf. Shindö 1983).

First, it can be seen as extreme cautiousness by the judicial authorities, concerned about the expanding social functions of anti-pollution lawsuits. In fact, in the wake of the Osaka Airport and Nagoya bullet train cases there was a rush to file lawsuits demanding injunctions against the pollution around or construction of large-scale public works and utilities such as airports, military bases, bullet trains, and thermal and nuclear power plants. There was an especially sharp rise in the number of lawsuits filed after the High Court decision in the Osaka Airport case that accepted almost all of the plaintiffs' demands (cf. note 6).

Thus judicial passivism must be seen, at least in part, to be an attempt to discourage further lawsuits by setting 'negative' precedents. As we have seen, in both the Osaka Airport and bullet train cases, the direct effects of granting the injunctions would have been quite limited. Judges in both the District and High Courts expressed sympathy for the defendant's concerns about a 'whole-line influence', a reference to the potential flow-on effects of a judicial decision. That is, for example, if the court had awarded an injunction for the JNR to slow down its bullet trains through a 7km residential stretch in Nagoya, people suffering from similar noise and vibrations elsewhere along the line could also demand that the trains reduce speed in their areas. On the basic principles of equity, the JNR would then have to concede to these demands. Thus accepting a few minutes delay in Nagoya would affect the entire Tökaidö line. The effect would be that what was once a 3 hour journey between Tokyo and Osaka at full operating speed would become a 6 hour trip or longer. But the judges' real concern seems to have been that creating a new precedent in one anti-pollution lawsuit would lead to a flood of similar lawsuits against other public works, public transport and general public works policies. In other words. they feared a 'domino effect' on public works in general as well as the entire bullet train network.

Second, if judicial injunctions were routinely awarded against public works, public utilities or public transport systems, judges would effectively be exercising authority over political decisions through their power to veto policy-matters such as the operating speed of bullet trains, allowable time-frames for airport operations, and whether to allow the construction of nuclear power stations or not. In other words, such decisions could violate the principle of the separation of powers between the executive, legislative and judicial branches of government. The judges are understandably reluctant to invite this. Judicial passivity towards anti-pollution cases is in part based on a feeling that 'these issues, that concern the State and its people as a whole, should be properly decided by the government or the Diet' (Matsuura 1980: 21).

Third, the judges of the Supreme Court are strongly inclined towards this view, and have the authority to overturn more 'progressive' lower court decisions. They also have the power to appoint, deploy and promote lower court judges; that is, they not only have the authority to judge the judges' decisions, but they *are* the authority that determines those judges' career prospects (Tsukahara 1990; Yukawa 1990).[12]

However, in the political situation in Japan, both then and now, the separation of powers has become almost perfunctory as a result of enlarged executive powers, slow legislative responses, an economically focused administrative bureaucracy, and a highly passive judiciary. There is insufficient debate about issues such as the extent to which the judiciary should be involved in political decisions, or the policy-forming function of the judiciary, and whether an alternative problem-solving process should be created. Judge Dantö's view emphasizing the importance and appropriateness of gradually reforming the laws through the accumulation of precedents—'the court, within the bounds of its original duty, should deal with new situations by devising creative interpretations and applications of the law' (Supreme Court Grand Court 1982: 58)—remained a minority opinion in the Supreme Court.

Lawsuits strategy and tasks for the residents' movements

As judicial negativism becomes entrenched, the effectiveness of the lawsuits strategy for anti-pollution movements will decline. The early lawsuits had ambivalent effects on the development of the particular residents' movement involved. Filing the suit strengthened the movement, but also produced new difficulties. These difficulties mount as the lawsuit drags on and judicial negativism increases.

Increasing reliance on lawyers can stagnate a movement

As a legal action drags on, a movement is at risk of becoming dependent upon its lawyers. The voluntary activities that once drove it may then stagnate. As explained above, the plaintiffs' counsel in anti-pollution lawsuits must assume a wide variety of roles. The plaintiff residents, who are typically laypersons in law, often find it difficult to understand the courtroom discussions or the prospects of their case. The plaintiffs' counsel must continually translate the legal proceedings into ordinary language for the plaintiffs during meetings to review each day's proceedings and through occasional information seminars and newsletters. Attendance at the hearings is typically rostered as an occasional obligation for most plaintiffs, when they must take time off work in order to sit through what is often mind-numbing tedium. Between the obscure legal jargon, the intricate interpretation of laws and the scientific and technical evidence required in an anti-pollution lawsuit, it is extremely difficult for plaintiff residents to take any initiative in these highly specific and technical trials.

Costs of the lawsuits strategy

Because the lawsuits strategy is generally only adopted as a 'last resort' after lobbying and protesting have proven insufficient to resolve the situation, although the slogans may claim that 'the movement and the trial are closely connected', it is natural for the plaintiffs' to become focused on the lawsuit once it has been filed, and increasingly so as the case proceeds. As time goes by it becomes ever more difficult to sustain the energy required to continue the previous activities of the residents' movements.

There are many demands on the plaintiffs' group during the trial. The plaintiffs are required to give evidence and formal damage statements, cooperate with site inspections and damage investigations, and attend the trial. The central leadership's burdens are especially heavy. They must maintain continuous contact with the lawyers, provide the information necessary for testimonies, make decisions about how to proceed, and attend to various business matters. They must also develop contacts and exchange information with other groups that are filing or engaged in anti-pollution lawsuits.

Also, as is well known, trials take a lot of time and money.[13] In anti-pollution lawsuits lawyers generally work on the case with no upfront payment. If the plaintiffs are victorious, the issue of fees is resolved when the defendant is ordered to bear the plaintiffs' costs over and above any compensation money. However, the continuation of the trial calls for other huge expenses such as photocopying evidence documents and reference materials, printing preparatory and other documents, transportation and remuneration of witnesses, several-days lodging for the defense counsel team, wages for the clerical staff etc. These expenses increase in proportion to the scale of the lawsuit and the number of plaintiffs, lawyers and witnesses.

Normally, the plaintiffs pay some proportion of counsel's expenses as the suit proceeds. The expenses born by the lawyers themselves may be substantial as well. If the plaintiffs' claims are totally dismissed, many of these expenses will not be paid. Extra money is sometimes collected from among the plaintiffs for events such as traveling to Tokyo for a national meeting or negotiations.

On top of these financial burdens, time commitments also accumulate as the trial progresses, such as attending hearings, debriefing meetings and information seminars. And these are merely the direct time burdens. If the suit drags on for, say, 10 years, the leaders will spend many years of their prime or advanced age in a litigation-centered life. Leadership responsibilities will force them to give up other life-plans they may have had. While the trial continues, their lives cannot be more then half-planned and provisional. Those who are tempted to drop the case and 'get on with their lives' are under additional pressure from their fellow plaintiffs and counsel, because each drop-out reduces the total number of charges being heard, the total compensation that might be awarded, and otherwise harms the prospects of the case.

Thus while concerns about their prospects of winning are ever-present for the plaintiffs from the time the suit is filed, as the case drags on and feelings of battle-weariness and fatigue grow, two other questions become increasingly prominent: 'how much longer will the case continue?' and 'how much will it cost?'

Throughout the formal legal proceedings, the plaintiffs are asked to refrain from activities outside the court that might lead to direct confrontation or negotiation with the defendant, as such actions may affect the outcome of the trial. The defendants typically also tend to avoid contact with the plaintiffs outside the court. Therefore, from the filing to the judgment, or until the parties agree to negotiate a settlement, contact between the plaintiffs (and other residents) and the defendant is generally indirect, mediated by the lawyers. During the trial period, there is typically a kind of deadlock between the two parties outside the court.

This tends to occlude most action by the non-plaintiff members of the original residents' group. There is typically then a corresponding tendency for the non-plaintiff members to withdraw from the movement group, exacerbating the difficulties of maintaining the movement's momentum.

The fluidity of the membership of voluntary residents' organizations has frequently been noted, but filing the lawsuit clarifies and formalizes the differences between the plaintiff and non-plaintiff members. Since non-plaintiffs are not likely to be entitled to any compensation awarded, the interests of the two groups begin to diverge. The energy of the movement tends to be increasingly concentrated on trial-related activities, and gradually the non-plaintiff members of the organizations become members in title only. Then there is a growing uneasiness about collecting membership fees from non-plaintiff members to use for legal expenses, and the organizations stop collecting these fees.

As the non-plaintiffs withdraw from the residents' group, the slogan that claimed that the 'residents' organization=plaintiffs' group' becomes a reality. Attempts to increase plaintiff numbers during successive lawsuits, as in the Osaka Airport lawsuit, may be possible when, as in that case, the problem arises from 'planar pollution' that affects many people over a wide area. More commonly, the plaintiffs' group becomes increasingly focused on 'defending the organization', especially from those members who may be tempted to abandon the case and withdraw the lawsuit.

Deviations between collective and individual interests

In disputes where pollution damages are still continuing, as in the bullet train and airport pollution cases, serious dilemmas arise around the acceptance of remedial measures (discussed in detail in Hasegawa and Hatanaka 1985: 189–97). If the defendants begin to offer partial measures or lesser sums in their efforts to resolve the matter out-of-court, as they typically do, a grim conflict invariably develops between individuals and groups of plaintiffs over whether to pursue an injunction ruling to realize collective objectives, or to accept, for example individual compensation for relocation and thereby alleviate one's personal and family's suffering. As the case drags on, rifts between the self-interested and the collectively- or principle-oriented participants widen. This tension inevitably saps energy from the residents' movement, irrespective of whether individuals accept compromises or the group's unity can be maintained. Occasionally splits occur between the self-interested and collectively-oriented groups. More commonly, individuals in the former category drop-out.

In the Nagoya bullet train case, the plaintiffs avoided any major schisms, but 147 actions were dropped, about a quarter of the initial number, either by plaintiffs who relocated or the families of deceased plaintiffs, before the conclusion of the initial trial (June 1979). This is perhaps the greatest difficulty in maintaining an extended lawsuit in the midst of daily, unending noise and vibrations.

After the initial trial, the plaintiffs' group changed their strategy direction. They increased their demands for concrete remedial measures through out-of-court negotiations with the JNR even while they were appealing against the ruling in their original lawsuit.

Remaining issues

Litigation also has a tendency to delay the resolution of problems that are not specifically listed on the complaint, or that occur later. In the collective actions characteristic of anti-pollution lawsuits, the lawsuit's claims are generally built around the lowest common denominator of all of the plaintiffs' individual damages. In other words, not all of the demands of the individual plaintiffs are included in the complaint. Typical 'hard to litigate' issues in the bullet train case included medical problems that required case by case responses, such as improving the recuperation environment for long term sufferers, and recovering hospitalization or health-related relocation expenses from the responsible party. At the same time it is difficult to negotiate resolutions to new problems that have arisen since the litigation began, such as measures to prevent or redress the desolation of the land vacated by residents who had relocated. Defendants typically assume a conservative posture towards any claims or matters outside the court, so there will be little if any progress in resolving such matters until the case is finalized.

Bottle neck of the lawsuits strategy

As we have seen, as the legal battle drags on the residents' movement that initiated the action and the plaintiffs' group that has carried it through both lose momentum. If the case continues for a prolonged period they invariably reach a point at which the only remaining outlet for exercising their power to resist is the trial—the only course of action available is to passively await the court's judgment. Everything else comes to a standstill until the judgment. And, if after all of that the judgment is unfavorable, it is much more difficult for the movement to explore new opportunities or avenues.

Thus a central question confronting environmental movements is how the lawsuits strategy can be utilized to increase the power of local residents' movements. In the mid 1980s, as the judiciary assumed a more passive position towards anti-pollution lawsuits, there was a growing need to reflect upon the 'lawsuits strategy'— to reconsider its possibilities and limitations, to evaluate its role and lessons, the results achieved and the issues remaining—if the environmental movement was to overcome this bottleneck.

8 Anti-nuclear power movements as new social movements

Preface

As of December 2003, with 52 commercial nuclear reactors generating electricity (excluding Monju and Fugen, non-commercial fast breeder reactors), Japan was the world's third largest producer of electricity from nuclear power plants, after the USA and France.[1] Following the partial meltdown in the No. 2 reactor-core at the Three Mile Island Nuclear Power Station in Pennsylvania, USA in March 1979, the explosion in the No. 4 reactor at the Chernobyl Nuclear Power Plant in the Ukraine, USSR in April 1986 strengthened the movements in many nations around the world that were calling for a change of direction—away from policies promoting the further development of nuclear energy. Yet still today, the Japanese government and others firmly maintain pro-nuclear energy policies. Among the citizenry, however, from around 1987 there was a large influx of women into the anti-nuclear movements, creating 'new wave anti-nuclear power movements'[2] throughout the country (Maruyama 1988; Nakajima 1988).

As we will see, since then the anti-nuclear movements in Japan have assumed the characteristics of the 'new social movements' defined by Touraine (1985), Melucci (1984), Offe (1985) and others. After briefly outlining the theories of 'new social movements' I will analyze developments in the Japanese anti-nuclear movements in the late 1980s through this theoretical framework. In the process I will explore the socio-political factors that underlie the widespread mobilization of women in the post-Chernobyl anti-nuclear movements in Japan.

Theories of 'new social movements'

The theories of 'new social movements' were developed primarily by European social movement researchers to explain the unique characteristics of movements that became prominent in advanced industrial societies from the 1960s, such as the environmental/

ecological, feminist, peace, and student movements. The formulation of the term 'new social movements' was intended to highlight their contrast to more conventional social movements, such as labor movements. The 'new social movements' theory developed by Melucci, Touraine and Offe has become a central perspective in the sociology of social movements, alongside the resource mobilization theory developed in the USA.[3]

Before examining the particular Japanese experience, I will briefly outline the four defining characteristics of the 'new social movements' identified by Offe (1985): (1) actor, (2) issue, (3) value, and (4) mode of action.

Actor: Periphery

The typical actors in new social movements are the peripheral citizens of modern industrial societies. They are the women, youth and ethnic minorities who have been excluded in one way or another from the modern ideals of 'freedom and equality', and who have been stripped of a self-definable identity. These identities, or at least the aspects of identity in question here, are not dependant upon their relation to the means of production, and are therefore not adequately (if at all) represented by the laborite framework of worker versus capitalist. Nor are such identities expressed through existing political classifications or social categories such as left and right, liberal and conservative, rich and poor, or rural and urban dwellers (Offe 1985). Rather, their identities have been socially defined by unchangeable, 'innate' factors such as gender, age and ethnicity.

These actors do of course have positions in the social structure— they have 'socio-economic status'. Except for ethnic minorities, both the primary activists and the less active supporters of new social movements are primarily from a new middle-class background.

Issues at the point of consumption

According to Offe, new social movements are involved in a wide range of issues, principally related to the life-world, neighborhood, cities, the physical environment, and the physical means of survival, and include local spaces for action, territory and body, health and sexual identity. Basically, they can be seen as issues at the point of consumption, in contrast to labor movements which generally fight over issues at the point of production. In other words, they reflect broader social changes in which the traditional boundary between the public (production) sphere and the private (consumption) sphere have faded away, as the private sphere has been gradually 'invaded' in a process that Habermas has called the 'colonization of the life-world' (1981: 37).

New social movements arose against the background of structural issues confronting modern society in the late 20th century, including socialism, the dominance of technocracy, and the rise of the welfare state. Running through all of these issues are the principles and ideals of civil society. Such principles are, however, fundamentally contradictory to the primary objectives—order and control—of the state. New social movements can therefore be seen as the actions of civil society to defend itself against the creeping extension of the state's techno-rational bureaucracy into the private sphere.

This increasing intervention of the ever-expanding rational-control structures is not only a problem in totalitarian or single-party regimes. In mature capitalist societies more complex and less tangible interventions are manifold. A prime example is the mass media's manipulation and control of desire. Less obvious examples include the development of gene manipulation and life management technologies. The rise of the welfare state radically extended the administrative powers of the state bureaucracy, in the process blurring the boundaries between the traditional public and private spheres. Among its results are an increasing dependency on the state, and a corresponding reduction in the autonomy of civil society.

Against this background, new social movements are strongly critical of the mass-consumption societies that have developed since World War II—societies that have prioritized economic growth (Hirsh 1988) and the welfare state (Offe 1985).

Changing value orientations: Autonomy and identity

The new social movements theorists have primarily focused on articulating the values and objectives underlying these movements. For Offe, Touraine, and Melucci, the basic values that connect this wide array of movements are autonomy and identity. More specifically, they include ensuring the decentralization of powers, self-governance, self-help, and self-determination. In Touraine's terms, the new social movements are, in short, struggles against compulsory or fixed lifestyles and technocratic social reforms, and struggles to defend the right to choose ones' own lifestyle and identity. Notably, the principle of self-determination here is anti-individualist, referring instead to a defense of collective identity. This is an important characteristic of the new social movements (Polletta and Jasper 2001). In a wider context, self-determination refers to a demand to defend the autonomy of the civil society from state encroachment, and demands for the 'actually existing' democratic system to become more genuinely democratic by including the social minority that has so far been excluded. More specifically, such demands may involve the management of production, the protection of a particular cultural identity, or the defense of a different lifestyle. Such objectives are not sought for their instrumental value in achieving other, supposedly 'higher' goals, but rather because they are expressive goods with their own unique values.

It is important to note however, that these new social movements characteristically limit their objectives to protecting civil society's autonomy and enlarging the public space and public sphere. They reject the illusions of wholesale social revolution or neo-Romantic communes. The other side of this is an implicit acceptance of the democratic state and the market economy, a characteristic Cohen calls 'self-limiting radicalism' (1985: 670).

Behavior patterns: Self-expression and networking

Regarding the behaviors that characterize the new social movements, theorists have focused on the non-everyday-life elements of social activism and the values thus expressed. According to Offe, strategies such as marching demonstrations, sit-ins and human chains are favored means of demonstrating that large numbers of people oppose the matter at hand. The objectives of these movements are typically defensive and obstructive, and their approach to achieving these objectives is typically intransigent—their demands are non-negotiable.

Structurally such movements are characterized by the absence or minimalism of formal organizations or groups. They commonly function as single-issue reactive type non-bureaucratic networks. They are typically averse to horizontal or vertical role-divisions within the movement group, and demand rigid adherence to non-bureaucratic direct democracy practices within the group.

Anti-nuclear movements: Before Chernobyl

Locally, in regional centers, and metropolitan areas

To what extent, then, do the Japanese anti-nuclear movements share the characteristics identified by the new social movements theories? This will be discussed focusing on four characteristics; the social position of women as the main bearers of the movements, their understanding of the issues, value orientation and mode of action. As previously noted (note 2), the phrase 'anti-nuclear movements' is used here in reference to those movements opposed to the civil use of nuclear energy, where the primary concerns are environmental. Although there is clearly some overlap between these and the anti-nuclear weapons movements, there are also significant differences and the two movements are clearly separate in Japan. This chapter does not discuss the anti-nuclear weapons movements.

The anti-nuclear movements in Japan can be divided into three categories dependant upon their geographical focus. First are local movements in the immediate vicinity of the site identified for a new power plant. Second are the movements in the regional centers, typically the capital city of the prefecture or other major cities within about 50km of the identified sites, with populations of 200,000 or more. Third are the movements in metropolitan areas, with populations in the millions, and no nuclear plants within the immediate or neighboring prefectures.

Using a case study[4] I conducted in 1988–89 and case specific materials about the anti-nuclear movements in various places, I will present an ideal-theoretical image of the movements prior to the Chernobyl accident, and compare this to the women-led citizen's movements after Chernobyl, with a view to establishing whether and in what ways Japanese anti-nuclear movements are 'new social movements'.

Local anti-construction movements

Local anti-nuclear movements typically arise in the immediate vicinity of a site selected for the construction of a nuclear power facility. These anti-construction movements are basically defensive and limited in nature. People in primary industries, such as fishermen and farmers whose livelihoods are threatened are the primary actors in these grass-roots movements. Their primary objective is to stop the construction and thereby protect their livelihood.

Entire families participate in these movements. Motivated by the defense of their livelihoods, such movements enjoy strong support by kin and geographical neighbors. The women's clubs organized by fishermen's and farmers' wives were frequently central forces in these movements.

Such women's groups—for example the 'Kattya Gundan' (Mums Corp: a nick name for the 'Mothers' group protecting children from nuclear fuel facilities' in Rokkasho)—were highly praised for the time and energy they invested in these movements. But within the movements, as in each household, these women continued to be treated as 'mere laborers' or 'fighting troops'. The decision-making in both areas was typically the privilege of the male 'head of each household, and especially the male heads of major kinship groups'.

The primary focal points of these local movements have generally been: negotiation over land acquisitions, the fishing cooperative's approval for conducting research in their local seas to study the impact of hot waste water from the nuclear plant on fishes and seaweeds, and negotiations to acquire fishing rights as compensation for damages. In the planning processes for such projects, the formal and substantial opportunities scheduled for local residents to negotiate with the power plant licensee are limited to these three points. Both public hearings and other formal opportunities have been criticized for being overly ritualistic, one-way and procedural.

Nevertheless, by standing firm in opposition to the project on any of these three points, fishermen and farmers can obstruct/delay the construction of a nuclear power station. With this as leverage, the local anti-construction movement can sustain solidarity and a strong resistance, provided they have the support of the local mayor and a majority of the local council. Under these circumstances, a few local movements have succeeded in forcing the rejection of plans to construct nuclear power plants in their neighborhoods, and others continue to obstruct/delay projects at the time of writing in 2003.[5]

In many cases however, the local government leadership decides in favor of the promised economic and social benefits of the proposed nuclear facility, such as national government subsidies provided under the laws to promote the construction of power plants and the income from local property taxes. With the electric power company's enormous resources for manipulating information and influencing local institutions and decision-makers, and the eventual exhaustion of resistance activists during long drawn-out disputes, the anti-nuclear activists can become the minority within, say, the local fishing cooperative, leading to a vote in favor of selling the fishing rights.

Once construction is finally underway the plant increasingly becomes a fait accompli, and the anti-construction movement is limited by the closure of political opportunity. Even if a lawsuit strategy has been initiated (as described in chapter 7), the local sense of being politically effective begins to diminish. Residents who continue to resist become an even smaller minority, and their power wanes.

Of course, there are differences in the particular matters disputed at each proposed site, but many of the nuclear power facilities that are in operation today came about through a process similar to the one just described. It typically takes more than 15 years from the time a new nuclear power plant is officially proposed to the local authorities until the commencement of the plants' operation, and of these, the entire construction and commissioning process takes only the last 5 years. The first 10 years are invariably engaged in negotiations with local residents and anti-nuclear movements, an indication of the depth of opposition and resistance to the construction of nuclear power plants in Japan.

Support movements in regional centers

In major regional centers in close proximity to proposed nuclear facilities the anti-nuclear movement is typically a support type movement. Prefectural and district level labor union organizations, affiliated either with the former General Council of Trade Unions of Japan (Sōhyō), the former Japan Socialist Party (Social Democratic Party) or the Japan Communist Party, are often the mobilizing forces behind these support groups in the first stages of a dispute. Other support groups emerge to offer technical and other specialist knowledge about nuclear power facilities to farmers, fishermen and local residents, sometimes in the form of general educational programs. Many of the central figures in these movements were politically socialized during the university unrest of the late 1960s, and came to be critical of big science and technology and the commercially focused research produced by industry-university collaborations. These activists have also played important roles in a lawsuit strategy, providing specialized technical knowledge from a perspective outside of the government-industry power loop. Their activities include testing the hot waste water and its impact from a power plant after it begins operations, closely monitoring the companies' responses to accidents or malfunctions, and protesting.

As the local resistance movements begin to stagnate, these anti-nuclear groups are generally the ones to continue the movement.

Notably, before Chernobyl, women's participation at this level was very rare, limited to a few professionals such as teachers or labor union officials. In most cases the women's perspective was not highly valued. Generally, participation at this level of social movement was limited to a few people. Although activists had long recognized the need to expand their groups' membership bases if they were to achieve their objectives, real progress was yet to be made.

Anti-nuclear movements in metropolitan areas

Anti-nuclear movements in metropolitan areas were typically driven by a core group of nuclear energy specialists who were not employed by either the government or industry. Specifically, groups such as the All Japan Anti-Nuclear Liaison Association and the Citizens' Nuclear Information Center acted as hubs in nation-wide information networks. These organizations supported local movements in the midst of disputes by giving lectures and conducting seminars or testifying as expert witnesses for the plaintiffs in a trial. Otherwise, their principal activities included: distributing information about nuclear power stations and other nuclear facilities, and about the status and details of the various anti-nuclear struggles around Japan; soliciting support for those local movements; publicizing moves by the government and electric companies' in these struggles, and keeping the Japanese anti-nuclear movements informed about the operation of nuclear facilities and the status of anti-nuclear struggles overseas (Takagi 1999). Until the early 1980s, women's participation was quite limited in this group.[6]

Anti-nuclear movements as 'new social movements'

After Chernobyl

An explosion in the No. 4 reactor at Chernobyl in 1986 released a huge cloud of radioactive dust that spread across the northern hemisphere. Coming in the wake of the partial meltdown of the No. 2 reactor at Three Mile Island in 1979, Chernobyl effectively made previously abstract concerns about the risk of serious accidents and the potential severity of global radioactive contamination a concrete reality, forcing ordinary people around the world to confront it at their dining tables as concerns about food contamination rapidly escalated.

In Japan, the public's new demand for information was fed by a plethora of articles, books and public lectures by activists such as Takashi Hirose and others. After 1987, the anti-nuclear movements spread like wildfire, becoming citizens' movements to protest nuclear energy on a scale never before seen. Many new citizen's groups were created in the process—primarily grass-roots-type movements that arose in the regional centers and metropolitan areas. The most distinctive characteristic of these new movements was that they were predominantly mobilized and driven by women.

Here the term 'movements' takes on a new distinctiveness. Whereas previously it could often be interchanged with the terms 'organizations' or 'groups', the latter terms would be misleading if applied here. These were individual-based networks independent of existing labor unions, political parties, and neighborhood-based organizations. To the extent that they were 'organized' it was on principles that rejected the norms of bureaucratic organization in favor of loose horizontal connections.

A good example can be found in the case of the proposed nuclear fuel processing and waste facilities in the village of Rokkasho, Aomori Prefecture. Citizen's groups in the regional cities such as Hachinohe, Hirosaki and Aomori created diverse networks between the Rokkasho residents, various residents' organizations in the surrounding areas (especially farmers' organizations), and activists in metropolitan areas. If the proposed facilities in Rokkasho proceeded as planned, it would become the largest (non-military) single-site radioactive materials handling facility in the world. The activists from these regional centers, with populations of 2–300,000, located between 50 and 100km from the proposed site, were motivated by what might be called a 'semi-local' awareness.

The opposition movement spread rapidly—from 1984 to 1990 newspaper opinion polls found that opposition to the proposed facilities increased from 35% to 62% in Aomori Prefecture, a large majority of the population.[7] The affects of this opposition were felt in the national Diet. Before 1989 the Liberal Democratic Party had occupied all nine of the upper and lower house seats from the Aomori electorates, but in 1989 the Socialist Party won an upper house seat, and in 1990 it won two lower house seats, one from each of Aomori's two electorates.

In another protest movement against trials to adjust the operational output level of the Ikata Nuclear Power Station in Ehime Prefecture,

4–5,000 people gathered from around the country to protest in a rally in the city of Takamatsu, where the Shikoku Electric Company is headquartered. This demonstration drew national attention as a 'new wave anti-nuclear power movement'. At the center of this appeal to the entire nation were the women of Beppu, a small city in Oita Prefecture about 70km across the sea from the Ikata plant. Rokkasho and Ikata are typical examples of post-Chernobyl anti-nuclear movements, in which women-centered citizen's movements in the regional centers provided connections between the local residents' movements and the citizens' movements in metropolitan areas. Thus through the women's networking, the anti-nuclear activism in response to proposed nuclear facilities spread nation-wide.

The next and largest event after the Ikata conflict was the 'Hibiya Rally to Stop the Nuclear Power Station' on 24 April 1988. More than 20,000 protestors joined this rally in Tokyo, more than four times the expected number. This was the largest protest activity in the history of the Japanese anti-nuclear energy movements.

Women as the carriers of the movement

Increasing political interest in citizens' and residents' movements, in general politics, and a growing awareness of themselves as political subjects are each factors in the social background against which women assumed the central roles in the anti-nuclear movements. Much of this change can be attributed to higher levels of education.

This trend is apparent in many areas of Japanese society. For example since the 1980s, there has been a steady increase in the number of women elected to local government offices. Most of these women had years of experience as local community leaders in various citizen's, residents' and consumer groups. The election of 22 women to the upper house of the Diet in the 1989 election, a phenomenon trivialized as a 'Madonna storm', was a landmark in the Japanese women's movements. This increasing public-political activism provides an important background for women's increasing involvement in the anti-nuclear movements.

Their increasing political and social activism encouraged more women to make independent political decisions, evidenced by the appearance of a 'gender gap' in public opinion polls. This gender gap is particularly apparent in attitudes towards nuclear power issues. For example, a national opinion poll conducted by the *Asahi Shimbun* in September 1988 found that 38% of men, but only 21% of women were in favor of further development of nuclear power generators, while

41% of men and 51% of women were opposed. A survey conducted by the Prime Minister's Office in August/September 1987 found that 67% of men and 47% of women agreed with increasing the nuclear power generating capacity. It appears that whether they participate in the movements or not, women generally have a stronger uneasiness or sense of pending crisis about nuclear power issues than men. However, further research is required to determine what background variables explain this gender gap on nuclear power issues.

Further research is also required to identify the demographic profiles of the women who were active in the anti-nuclear movements of the late 1980s and early 1990s. No specific quantitative data exists on these women, but we can approximate their demographics from surveys by Mototaka Mori (1988: referred to hereafter as 'the Ikego report')[8] of the activists involved in the Ikego US military housing construction dispute and the activists of the citizen's movement in Musashino City by Akihiko Takada (1990), plus various reference materials concerning the anti-nuclear movements, including my own interviews and observations.

While their ages range from 20 to 50 and beyond, the core of these movements is the cohort born during the first baby boom after the war (1947–49)—in their early 40s in 1989—while the 1945–54 cohort provided the driving force. For the mothers in this cohort, the post-1987 anti-nuclear movements took-off just as or after their youngest child reached school age. This is an important 'life-changing' event—a time when they must 'redefine their identity'.

Beyond the general effects of this life event, this cohort is also part of a rather unique generation. In childhood they experienced dramatic changes to their living environments during the high economic growth period that began in Japan in the late 1950s. In their late teens and early 20s they experienced the university unrest of the late 1960s and the first wave of anti-pollution movements in the late 1960s and early 1970s. Many of them became housewives in the late 1970s and early 1980s, when the roles and social status of women were being critically reviewed (1976–85 was the 'United Nations Decade for Women'). In short, this cohort of women grew up, went to school, got married and had children during a time of continuous social change. They were eye witnesses to the rapid industrialization and high economic growth that became known as the 'Japanese miracle', and the all-encompassing transformation from an industrializing society to a mass-consumer society. They are the last generation to know what life was like in Japan before this incredible social transformation began.

According to the 'Ikego report' mentioned above, up to one-third of the 'women activists' involved in that protest movement (48 respondents) reported previous participation in political activities or student movements. This suggests some correlation between political socialization experiences in youth and later participation in citizen's movements.[9]

This cohort of women is also more highly educated than any previous generation of women in Japanese history.[10] The Ikego report cited a survey conducted for general adult women in Zushi City which found that 15% were graduates of a 4-year university or higher, while among the women activists involved in the Ikego protest movement, 58% were in this category. When junior colleges are included, this becomes 33% of the general female population, and 77% of the women activists. A similar trend would be found among the activists of the anti-nuclear movements. In short, the women of the first baby boom and their 1945–54 cohort were the first women to receive higher education en masse in Japan, and as they completed their qualifications in the mid 1970s they began to have an influence in local areas of Japanese society.

But there is also strong evidence to suggest that at least half of the women actively involved in the anti-nuclear movements are full-time housewives. A survey of activists in Musashino found that 64% of the women activists were full-time housewives. A similar survey in Zushi found 56%. The latter also found that 17% were self-employed. Studies of local residents' movements have frequently noted the importance of 'full-time citizens', full-time housewives, self-employed and elderly people who spend 24 hours a day in the local area, and have time resources available. This demographic also plays a central role in the anti-nuclear movements.

As we have seen, the women at the center of the anti-nuclear movements in the regional centers and metropolitan areas are typically full-time housewives or self-employed persons, are more likely to have higher education qualifications than not, and more or less center on the cohort in their late 30s and early 40s in 1989. The anti-nuclear energy movements of the 1990s continued to be carried by the women of the 1945–54 cohort. Younger cohorts did not join in large numbers.

It appears that for this generation of women, their identities as mothers began to waver when their children started school, and they redefined themselves as mothers through social participation in the anti-nuclear movements. That is, many created new self-identities as 'mothers protecting their children from radiation'.

Nuclear energy issues: The housewives' and mothers' perspective

As noted, women were clearly involved in the anti-nuclear movements
in Japan before Chernobyl, but except for the local residents'
movements, the anti-nuclear movements were generally considered
to be be 'men's issues'. Although they may have had concerns, very
few women actively committed to the movements at the regional or
urban levels.

Since the 1970s, housewives have become increasingly interested
in health and environmental issues, especially those concerning food,
such as pesticides and other contamination. In fact, everyday life
conditions, such as the sharp rise in the incidents of atopic dermatitis
in young children, meant that mothers could not avoid being sensitive
about the relationships between the environment, food supplies,
and their own and their children's health. This is an important
background for a potential interest in the safety of nuclear energy and
its health effects. The Chernobyl accident directly connected fears
of radioactive contamination to the dining table. The most mundane
and routine of the housewives' chores—choosing and preparing safe
and nutritious food for her family—became fraught with the risk
of radioactive contamination and made these mothers immediately
aware that they and their children are direct stakeholders in nuclear
energy issues.

Various books on the issue reinforced this new awareness. For
example, Takashi Hirose (1987) begins his book warning about
the risks of a severe accident in a nuclear power plant in Japan by
acknowledging his personal position as a father of two daughters.
Taeko Kansha (1987) stresses that her involvement in the effort to
'stop the nuclear power station' was an extension of her responsibilities
as a mother to 'protect the lives of her children'. These messages were
very influential in Japan at the time, especially among housewives.

In Japan, the words 'mother' and 'parent' signify traditional
notions that serve to justify women's activism and were employed
to great strategic effect in mobilizing women into various social
movements. They signify roles and values that were central in these
movements' orienting frameworks. This is clearly evident in the
activities of the Japan Mothers' Conference (Nihon Hahaoya Taikai),
movements against nuclear weapons, local residents' movements
opposing new military facilities and the broader peace movements.

For the actors themselves, the role of 'mother/parent' is already
central to their orienting framework, a key determinant in their

behavior patterns and decision-making. Appealing to someone's sense of responsibility 'as mother/parent' can be a very direct and effective means of motivating them to action. As one mother put it, 'Although it may sound self-centered, when I first heard about [the issue], the first thing that came to my mind was that above everything else, I didn't want my child to get exposed to such radiation.'[11] From this first identification of the social problem comes further interpretation, efforts to identify and verify culpability; then anger and communicating one's feeling or sense of crisis to others, which intensifies the motivation to participate and helps to overcome the inertia of helplessness. This series of processes is strongly shaped by the cultural framework: 'mother/parent'.

More broadly, in the Japanese socio-political context, where the normative value of 'citizen' and 'civil society' is rather weak, the iconic 'mother/parent' can signify the universality and solidarity of the movement. It has come to symbolize a 'post-political' responsibility for others beyond the particular parent/child relationship, and can thereby position a social movement as above or outside of ideology, political parties and self-interests.

At another level, the idealized 'mother' signifies a role in which women and 'women's knowledge' is socially valued, and is thus an especially effective symbol for social movements. In a society where a woman's comments are typically not valued, and are often trivialized as merely 'the idle chatter of women and children', at the same time, the 'mother's word' or 'mother's perspective' is highly valued, representing a devoted, self-sacrificing and altruistic position.

The mass media's stereotyped portrayals of the anti-nuclear activists as 'ordinary housewives' and 'ordinary mothers' further demonstrates the dual nature of this iconic social role/identity. The activists themselves are well aware of the 'catchy' labels in the mass media, as well as the negative stereotypes that accompany them.[12] Nevertheless, if somewhat paradoxically, the activists' identities as 'mothers' have been used as a justification for women, and especially housewives, coming out of their private spaces to become social activists in the public sphere.

Values of the anti-nuclear movements: Defending self-determination and self-expression

'What is demanded of the anti-nuclear power movement now is not "how to abolish the nuclear power stations", but "how to counter them with the existing anti-nuclear movements".'[13] This statement plainly describes the value orientation of the post-Chernobyl anti-nuclear energy movements.

Ryōko Obara, a 40 year old woman from Beppu City at the center of the protest movement against the trial operation to adjust the operational output level at the Ikata power station, described her involvement as follows. 'The nuclear power stations were supported by people's ignorance, including myself, who didn't know and allowed today's situation to occur. When I became aware of this, I thought I had to *take back my free will*, and *regain my voice*, and stop the nuclear power station *as a way of expressing myself*' (Obara 1988: 22, emphasis added).

The keyword here is 'myself'. What is being questioned is the self who has enjoyed the benefits of the mass consumption culture and advanced industrial society, and the lifestyle of this self who once believed in science and technology and the mass media's slant on the 'news' and was ignorant of the 'terrors' of nuclear power. Here we see a simple and straightforward connection between one's individual life/identity and the modern civilization that produced and is dependent upon nuclear power stations. The challenges of changing one's lifestyle—of redefining one's self-identity and strategically expressing one's objections to the nuclear power stations through the protest movement come together here unmediated. The central task of the movement, then, is to defend self-determination and self-expression. This orientation is typical of the post-Chernobyl anti-nuclear movements, generating a vast array of single-issue grass-roots-type networks in the regional centers and metropolitan areas.

Correspondence between the movements' goals and personal feelings

Motivated by values such as self-determination and self-expression, the correspondence between one's personal feelings and the movements' goals is of the utmost importance. Ryōko Obara, cited above, recalls the excitement in the barricades during the Ikata conflict.

> We really expressed our will, with all our power. All of us were wondering how to express our anger, and in the end we did it through songs and dances. At first, we were just chanting 'stop the nuclear power stations'—not much of a song. But it gradually became longer

and longer and a rhythm developed, and it became music. In the end we were singing a song accompanied by percussion... It wasn't our objective in the beginning but it was fun once we started. Power just kept flowing like a spring from within. (Obara 1988: 28)

At the time various anti-nuclear movement meetings were aiming for 'performance-based, informal rallies', and were characterized by the 'wide, flexible and collective expression of individual will, through a combination of songs, dances, speeches, zechin-cloths and flags' (Maruyama 1988: 64). This approach entailed an implicit critique of the more conventional mass mobilization type of rally typically organized by labor unions. The aim was to reduce the control of the organizers and leaders as much as possible, as such control would obstruct the individual participants' freedom of self-expression and spontaneous release of energy.

The tendency towards festival-like demonstrations and the radical rejection of controlling processes is prominent in the movements of western Japan and the Tokyo area, but less so in the movements of northeast Japan. Even here, though, the quest for a style of action that corresponds to the activists' alternative values and aspirations is evident in the names of protest groups and their newsletters, which express creativity and originality. For example, nature is evoked in names such as the 'Apple blossom group' (Aomori Prefecture) and 'Grapes anti-nuclear energy group' (Miyagi Prefecture), while slogans like 'Stop the construction of nuclear fuel facilities!' (Kakunen maine!) and 'Stop the construction of the Onagawa No. 2 reactor' (Odazunayo genpatsu zousetsu) using local dialects for nuance,[14] and newsletter names such as 'Energetic housewives newsletter' (Hirosaki City) and 'Post-nuclear power petition campaign—Wakuwaku newspaper' (Miyagi Prefecture) identify with their readers or their shared objectives.

Anti-nuclear sentiments are expressed in many ways. A guide-book for anti-nuclear movements (Yūki 1988), based on the activities of a housewife-centered group in Kunitachi City, was produced because 'after hosting a public lecture about nuclear power issues, enquiries kept coming in saying they also wanted to declare an anti-nuclear position, and *asking for details on how they should go about it*' (Yūki 1988: 105, emphasis added). The book shares practical knowledge on a range of topics such as how to recruit members, organizing study activities, editing newsletters, distributing handouts, collecting signatures, holding public lectures,

collective action to pay electricity bills late, protesting to the
electricity companies through telephone campaigns, writing to
newspapers and magazines, participating in rallies, measuring the
residual radioactivity in foodstuffs, participating in radiation-watch
networks using simple radiation detectors (R-DAN), boycotting the
products of nuclear power related companies, supporting the local
(on-site) anti-nuclear movements and fundraising.

Signing petitions also acquired new meaning as a protest activity—
as a means of self expression. Numerous petition campaigns have
been conducted, such as the 1989–90 Petition for a Denuclearization
Law which collected almost 3.3 million signatures. Having relatively
low participation costs and risks, a petition is a relatively easy means
of inviting participation, giving individuals an opportunity to voice
their opposition to a particular issue in their own names.

Networks as the organizing principle of groups that embody such values

The women-centered anti-nuclear movements of recent years
have been organized as loose and horizontal connections, roughly
corresponding to Lipnack's definition of a 'network' (Lipnack and
Stamps 1982). Summarizing the key characteristics of networks, we

Figure 8.1 Network structure of anti-nuclear energy movements

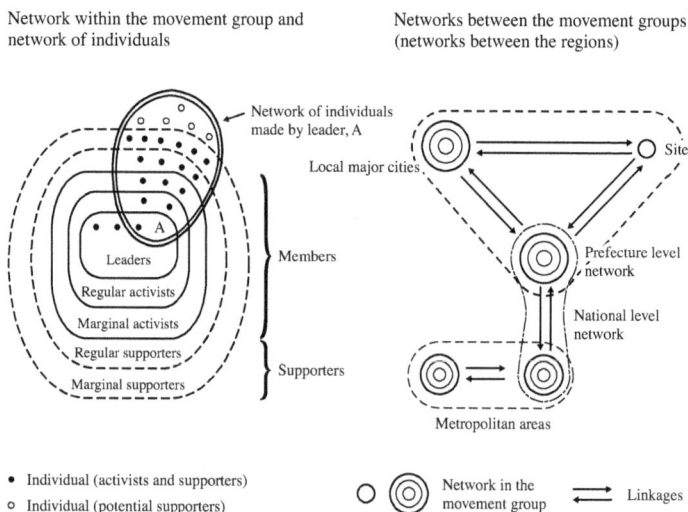

find that they: 1. have a non-hierarchical internal structure, 2. are decentralized, 3. consist of components that are independent of each other, 4. have ambiguous boundaries, 5. are linked by shared values and goals, 6. value the processes of social relationships themselves rather than their instrumental efficacy, and 7. are channels for mobilizing resources and communication (refer to Figure 8.1). In such networks, the line dividing members and non-members—the included and excluded—are vague. An individual's participation in the movements' activities is voluntary, and whether one only occasionally attends a rally or is a highly active organizer of activities, each contribution is valued. Hence, supported by the day-to-day relationships of the housewives' networks, there is great potential for mobilizing resources.

For example, a fundraising advertisement with the tag-line 'Stop the nuclear fuel facilities (Kakunen maine)!' was printed in the local newspaper and three others on 9 April 1987 by the 'Mothers group protecting children from radiation'. The campaign raised a total 2.8 million yen in donations from about 1,500 people across Aomori Prefecture and throughout Japan.[15] Another advertising campaign under the banner 'Nuclear power generation—You don't need it, do you?!' (Iranaissho!) appeared from 26 October 1987 in four newspapers in Hokkaidō. This was an appeal begun by six housewives from Asahikawa City attempting to stop the construction of the Tomari Nuclear Power Station and stop plans for a high level radioactive-waste facility in Horonobe. The campaign attracted 6.25 million yen in donations from 4,000 people, mainly from Hokkaidō (All Hokkaidō Opinion Advertising Coalition 1988: 93–109).

Closing remarks

Social movements generally face a series of dilemmas around decisions about goals and strategies. These dilemmas can be characterized by the following set of dichotomies: 1. interests or values, 2. means or expression, 3. social transformation or self-transformation, and 4. organization or network (Hasegawa 1990). 'New social movements' tend to emphasize the latter in each of these fields.

To contrast these diagrammatically, the movements before Chernobyl were typically local anti-construction movements organized and driven by self-interested farmers and fishermen with support from other local residents and groups in the neighboring cities, for whom the movement was a means towards social transformation (or in this case, its inverse—conservation). After Chernobyl the movements were primarily carried by women living in the regional centers or metropolitan areas, largely in accordance with the principles of the 'new social movements': values, expression, self-transformation and networking.

As we have seen, a typical profile of these post-Chernobyl social actors might describe a woman who had children starting or about to start school in 1987–90 and therefore needing to redefine her identity, with a relatively high level of education, and living in an urban environment. These identity-oriented movements were inherently limited in their capacity to achieve concrete objectives. Although the overall goal of the movements was to turn Japanese society away from its nuclear power-dependant policies, the anti-nuclear movements in Japan have had little if any impact on the political process. In contrast to Germany and the Scandinavian countries, in spite of tremendous growth in the late 1980s and early 1990s the Japanese movements have never been large enough to have a significant impact in national elections. As we enter the 21st century, the anti-nuclear power movements in Japan still do not have enough influence to pressure the government or electricity companies to change their policies or practices.

In 1991, I concluded the first publication of this argument with the final paragraph,

> Will the anti-nuclear energy movements with women as its new bearers remain a temporary movement post-Chernobyl, or will it continue to expand? It seems the anti-nuclear energy movements are being tested in their political effectiveness and their degree of influence on political processes. (Hasegawa 1991b: 56–7)

At the time, I was afraid that this self-oriented activism would be temporary and fade away. This fear was realized in the early 1990s. As the shock of Chernobyl receded, the activities of the movements called the 'new wave anti-nuclear power movements' rapidly declined. The anti-nuclear power movements became active again from the autumn of 1994 to August 1996, with a movement calling for a public referendum to decide whether to build a nuclear power station in Maki Machi, Niigata Prefecture or not. But the heat of the mass protest activities remained limited to this town and the neighboring areas; it did not spread nation wide. This is the subject of the next chapter.

9 Regional referendums: Community responses to nuclear facilities

Introduction

In the 1990s, Japan saw a surge in the number of regional referendums held on questions about nuclear power facilities and other environmental issues, and an even stronger surge in unmet demand for such referendums. In this chapter I will examine the sociological background to this new demand, the significance of referendums for social movements, and the conditions necessary for their success. The public's increasing demand for self-determination, difficulties in obtaining court injunctions (see chapter 7) and the decline of the labor movement have all contributed to shifting the main focus of citizens' movements to the legislative arena. Organizing a referendum helps a social movement to mature, to refine their strategies and objectives, and garner public support through energetic campaigning. In the process, residents learn a great deal about civil society, and thereby enhance representative democracy.

This chapter will proceed primarily through an analysis of the referendum that was held in Maki Machi, Niigata Prefecture, in August 1996 and the unsuccessful opposition to the proposed nuclear fuel processing facilities in Rokkasho Mura, Aomori Prefecture, focusing on the differences in their political opportunities, mobilization structures, and cultural frames. Maki Machi, neighboring the prefectural capital and with many urban professional residents, was not an isolated or marginalized village. There were two key turning points for the Maki residents: the formation of a group demanding an official referendum, and the unofficial referendum this group held, in which almost half of the eligible voters participated. Through cultural activities such as collecting folded origami cranes and creating 'trees of happiness' from decorated handkerchiefs, the campaign for a referendum was a grassroots movement opposing the planned nuclear power plant and demanding self-determination. Rokkasho Mura, in contrast, is in a quite marginalized region and, although in the early stages (pre-1986) local fishermen in the village were the key players, after the second stage the local opposition movement was far more

dependent upon supporters and sympathizers from the regional cities
and metropolitan areas.

The role of referendums

The impact of the country's first referendum

On 4 August 1996 in Maki Machi, Niigata Prefecture, the populace
voted on the planned nuclear power plant in the nation's first referendum based on a local ordinance.[1] The turn-out was 88.3% of eligible
voters, with 60.9% voting 'No'. Although the results of the referendum were not legally binding, this clearly marked the end of this
particular nuclear power plant project. Tōhoku Electric Company announced their abandonment of this project on 24 December 2003, as
discussed shortly. The referendum results proved to be the first direct
step towards this decision. The referendum in Maki Machi sparked
a series of referendums in other parts of the country. Between the
Maki Machi referendum and the end of 2003, there were 12 officially sanctioned referendums in Japan: three involved nuclear power
plants, two concerned US military bases, five involved industrial
waste dumps, one was about the construction of a weir, and the other
was about a stone quarry. Many other calls for regional legislatures
to sanction referendums have been rejected, but attempting to hold a
referendum has nevertheless become a principal strategy for social
movements since the mid 1990s. Numerous residents' opposition
movements have received wide publicity in Japan from reports in the
mainstream media about their demands for referendums concerning
large public works projects.

Referendums have become a significant mode of gauging residents'
opinions about the introduction of dangerous or nuisance facilities
and public works that may have detrimental environmental effects.
Referendums have generally been initiated by residents who oppose
the proposal, but some have been organized by the elected leaders of
local authorities, such as the referendum concerning the proposal for
an industrial waste dump in Shiroishi City, Miyagi Prefecture (Imai
ed. 1997; Hasegawa 1998f). Another referendum was called for in
Miyama Chō, Mie Prefecture, by the proponents of a nuclear power
plant.

Kubokawa Chō, Kōchi, was the first municipality to proclaim a
local ordinance enabling a referendum on a nuclear power plant on
22 July 1982. Nantō Chō, Mie, followed on 26 February 1993 with

ordinances to enable a referendum about a proposed nuclear power plant and again on 24 March 1995 about a preparatory environmental impact study. Other local ordinances soon followed: Kushima City, Miyazaki on 8 October 1993, Maki Machi, Niigata on 3 October 1995, Kisei Machi, Mie on 14 December 1995, Kariwa Mura, Niigata on 25 April 2001 and Miyama Chö, Mie on 27 September 2001. Of the ordinances cited here, only three actually proceeded to a referendum, including the one in Maki Machi. On 27 May 2001 the people of Kariwa rejected plans to convert an existing nuclear power plant to MOX fuel (the plutonium-thermal project). The residents of Miyama rejected a proposed nuclear power plant on 18 November 2001.

At Kariwa Mura, where seven massive reactors were in operation, 53.6% of voters said 'No' to the proposed MOX fuel conversion. The operator, Tokyo Electric Power (TEPCO), had planned to commence MOX operations in 1999, but had to postpone until 2002 because of the referendum. In August 2002, systematic cover-ups and falsification of data about malfunctions and accidents over an extended period became public knowledge and the company was strongly criticized by local governments, citizens and the media. It is now extremely uncertain if or when the MOX program will start. A similar MOX program planned by another giant nuclear power company, Kansai Electric Power (KEPCO), has also stalled due to a data falsification scandal.

Miyama Chö's referendum was organized by local business people who thought a nuclear reactor would boost the local economy, which was suffering from a decline in primary industry. When 67.5% of the town's eligible voters said 'No', the Mayor declared the proposal dead.

In all three referendums concerning nuclear power proposals, clear majorities have opposed the projects and successfully prevented further development. On other issues, the referendum about an enormous new weir across the Yoshino River in Tokushima effectively put an end to the project. Similarly, a proposed industrial waste dump in Shiroishi was abandoned by the company after the prefecture's governor rejected it in respect of the referendum's results. (A subsequent lawsuit by the company seeking compensation was lost.) Although regional referendums are not legally binding, their political significance cannot be ignored. That is, they are to some extent 'politically binding' for the elected leaders of regional authorities.

The social and political background

The right to hold a referendum was officially enshrined in the Local Autonomy Law, enacted in 1947. A sociological analysis is therefore required to explain the sudden upsurge in their usage. Indeed, sociologists have recently produced a number of works on this subject.[2] From my own case studies and empirical observations of Maki Machi and Shiroishi, the following factors can be identified.

First, there was clearly a growing demand for self-determination throughout Japan. In a representative democracy where voters occasionally elect representatives to make decisions on their behalf, referendums provide a means through which the electorate can directly express its opinion on a particular issue. The system is not perfect, since various residents (such as foreign citizens) are excluded from voting, but it is nevertheless open to all eligible voters and thus a relatively direct form of democratic decision-making. Although a regional referendum is not legally binding, it can be held if the local ordinance to implement it is passed in the local council. The demand for self-determination and a greater voice does not, at present, show any signs of receding, which suggests that the demand for referendums will continue to intensify on a broad range of issues.

A second factor moving residents towards referendums is the extreme difficulty in obtaining court-ordered injunctions in environmental lawsuits. Although the judiciary has played an important role in alleviating the suffering of victims through cash compensation awards after the fact, as for example in the four major anti-pollution cases, it has been reluctant to suspend or intervene in public works and facilities, as in the 1981 Supreme Court verdict on the Osaka Airport. Furthermore, because litigation can take a very long time, activists tend to become preoccupied with legal matters, and the campaign typically ends up being led by the legal team (see chapter 7).

A third factor is the limited power of regional bodies. Even though the 'decentralization of power' is a fashionable concept at the moment, more often than not the regional bodies have very limited capacity and resources due to highly centralized financial and administrative structures. With the paths to judicial and executive remedies to social issues seen as 'closed', opposition movements have taken the path towards the remaining possibility, legislative power. The regional legislative bodies are generally controlled by conservative groups with economic interests who are eager to invite development projects and reluctant to listen to the environmental concerns of citizens. These groups also enjoy strong connections with the politicians of the prefectural assemblies and Diet (see Broadbent 1998). The regional legislative bodies are typically the greatest barrier to citizens' efforts to hold a referendum. Where this barrier has been overcome, referendums have proven to be an effective vehicle for influencing political decisions.

A fourth factor is the decline of the organized labor movement, especially the sectors affiliated with the former Japan Socialist Party (now Social Democratic Party). The strategy of mass mobilization through traditional organizational structures is becoming increasingly ineffective, as are directly confrontational campaigns (pickets, strikes, boycotts etc). The referendum is attractive, in contrast, because it allows individual citizens to determine their own level of participation, relatively free of the often burdensome trade union and political party 'machines'.

What this means for social movements

Initiating and pursuing a referendum has numerous affects for the social movement itself. First, in the campaign for a referendum, objectives, strategies, time-schedules and issues become clearly defined. Second, to initiate a referendum requires broad support from the general public. It is therefore essential to build a grand coalition of people from across the political spectrum. People of the 'left', for example, must set aside their political differences to join with conservatives and minorities in a campaign over a single issue. Once established, this coalition must be sustained until voting day. Initiating a referendum is thus in many ways an avenue for a social movement to mature.

Third, at the same time, a referendum is a learning opportunity for the residents. I was involved in conducting an exit poll survey during a referendum in Shiroishi in June 1998, which confirmed that the process leading up to the referendum had had a tremendously stimulating affect on the voters (Hasegawa 1998f). In response to a question about what they had done to prepare for the referendum, nearly half of the respondents (49.4%) said that they had discussed the issue with friends and family. A similar proportion (46.2%) reported that they had started trying to reduce their household waste and to recycle more, 34.8% had consciously sought additional information about the issue, while 27.1% had urged friends and acquaintances to vote. When asked about other environmental issues, 94.7% said they were concerned about dioxin pollution caused by burning waste, 82.6% about forest preservation, and 81.8% about global warming. In short, the referendum was clearly a learning experience for the residents. It inspired them to seek more information, and as they became more knowledgeable they discussed the issue with friends, family and acquaintances, while at the same time altering their lifestyles to reduce their own waste production and recycle more.

Returning the ball

Conservatives often criticize regional referendums for being 'an affront to parliamentary democracy', in that they are not legally binding and are typically expressions of selfish nimbyism (Not In My Back Yard-ism). National policy making, they argue, should not be decided by narrow regionalism. Let us briefly examine these criticisms.

To hold a regional referendum in Japan, a local ordinance or a prefectural law must be carried by a majority of the relevant elected assembly. This fundamental condition means that a referendum is in fact part of a representative democratic system. Indeed, during the huge upsurge in referendums during the 1990s, more often than not even when a majority of the residents were in favor of a referendum, the relevant elected authority did not pass the necessary legislation.

In other words, referendums augment representative democracy. To make it happen, the endorsement and commitment of the local political leadership is crucial, as was demonstrated in Maki Machi, Okinawa Prefecture and Shiroishi City. In the Okinawa referendum held on 8 September 1996, just one month after Maki Machi's, on reducing the size and activities on the US military base—the first and so far the only referendum at prefecture level in Japan—Governor Öta assumed a crucial leadership role for implementing the referendum (see Imai 2000). Takaaki Sasaguchi, the Mayor of Maki Machi, declared on the day of their referendum that even though it was not legally binding, the result of the referendum would be honored and his future decision about the sale of the publicly owned land in the middle of the proposed reactor site would be determined by the voters' will. It would bind the town politically in the future, he said. Leadership like this is indispensable in raising the consciousness of the residents and in making them realize the importance of their vote. Conversely, the higher the voter turn-out, the harder it is to ignore the referendum's results.

While particular cases of opposition to 'dangerous facilities' or similar may justifiably be characterized as selfish or nimbyish, it is equally as clear that citizens have a right to protect their local environment from having such facilities imposed upon them. This right is the fundamental basis of regional autonomy. It is therefore not justified to accuse the local residents of being selfish nimbys unless all necessary procedures and precautions have been properly carried out. These must include detailed and critical examinations of the policies that affect the residents, residents' participation in decision-making processes, freedom of information, and thorough environmental impact studies that genuinely consider alternatives. When those who accuse local residents of being selfish and criticize regional referendums as nimbyism are the residents of distant cities, especially in metropolitan areas who will benefit from the new nuclear power facilities, or the business interests behind a development proposal, or the industrialists in desperate need of new waste dumps because there is nowhere left in the cities—we really must wonder who is calling whom 'selfish'.

Holding a referendum can be likened to throwing the ball back to the elected officials, requiring them to review and alter their policies in accordance with the expressed will of the people.

Factors behind the success of the Maki Machi referendum

The tide turns

To better understand the conditions for a successful referendum campaign I will analyze the successful Maki Machi anti-nuclear power plant referendum. Comparing this case with the unsuccessful movement opposing the nuclear fuel processing facilities at Rokkasho Mura in Aomori Prefecture will assist in highlighting significant variables (Funabashi et al. eds, 1998). This discussion is organized by a three-fold analytical framework—political opportunities, mobilization structures, and cultural framing (see chapter 5; McAdam et al. eds, 1996).

At the center of the Maki Machi dispute was a plan to build an 825MW boiling water reactor nearby. Maki Machi is the nearest neighboring town southwest of the prefectural capital city, Niigata (see Figure 9.1). The plan was formally announced by Tōhoku Electric Power in 1971, after first being reported in the local *Niigata Nippö* newspaper in June 1969, and approved by the national government's Power Development Council in November 1981. The Maki No. 1 reactor was thereafter included in national power development plans and policies, but the mandatory safety assessment was suspended in 1983, as the necessary land for the site had not been acquired. The Maki No. 1 reactor thus became the first officially approved reactor project in Japan to have completely stalled and to be finally abandoned.[3]

Figure 9.1   Maki Machi and the site of the planned nuclear power plant

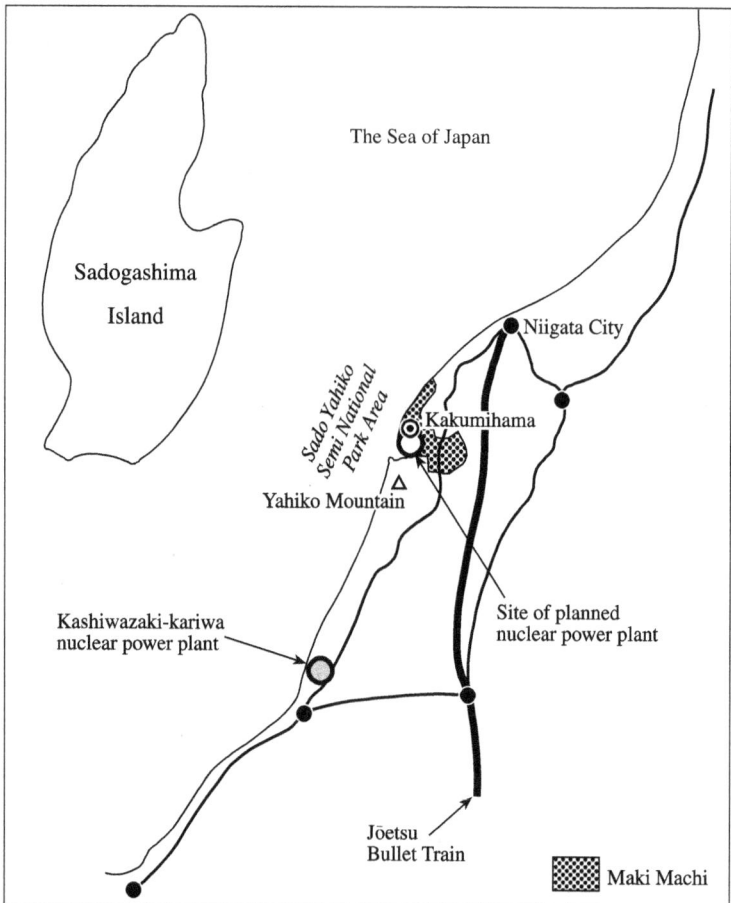

Note: adapted from Imai eds. (1997: 106)

As mentioned, the voter participation rate for the Maki Machi referendum was almost 90%, with more than 60% of them opposing the planned reactor. True to his word, the town's Mayor Sasaguchi honored the result, and refused to sell the 9,070.74m² of town land at the center of the proposed reactor site to the electric company. Sasaguchi also predicted that this referendum result would constrain mayoral decision-making in Maki Machi for at least twenty years. The Power Development Coordination office was abolished from the town hall in March 1997.

Proponents of the reactor subsequently turned their attention to the mayoral election scheduled for January 2001. If their candidate won he would then sell the town land in question to the electric company so the project could proceed. The majority of the town council was seized by conservatives and came to favor the proposal. Fearing this outcome, Sasaguchi decided to 'cement the residents' will as expressed in the referendum and end the issue' by selling a portion (743m²) of the town land in question to 23 members of the 'Group to bring about a local referendum' on 30 August 1999. The sale was finalized, secretly, on condition that the land would not be on-sold or leased to any third party. This residents' group in essence formed a trust to safeguard their town from the nuclear power plant. The sale of the land made it impossible to proceed with the Maki No. 1 reactor proposal, but neither Tōhoku Electric Power nor the national government officially abandoned the project at that time. The project was finally abandoned, as mentioned, in late 2003, after the proponents' lawsuit attempting to nullify this land sale was defeated in the District, High and Supreme Courts. Thus, by continuing their oppositional campaign through a referendum, Maki Machi became a prime example of a successful residents' opposition movement.

Why was it successful?

The August 1994 mayoral election had returned the incumbent, Kanji Satō, for a third term. Satō was very eager to push ahead with the power plant and until April 1995, like most other towns, only one or two of Maki Machi's 22 elected councilors clearly opposed the nuclear development project. In other words, the local residents opposing the power plant faced the same difficulties experienced in other areas. So, although the mass media tends to portray Maki Machi's success as a foregone conclusion, a sociological analysis is required to explain how the tide turned so dramatically in such a short time.

The biggest turning point in the resistance movement was the formation of the 'Group to bring about a local referendum' in October 1994. The group was formed by residents who had not previously been involved in the anti-nuclear movement in this town. It opted for a different approach and won the support of many of the people who the existing movement had failed to attract. The group staged an independent (unofficial) referendum, in early 1995. With a turnout of 45% of eligible voters, and 95% of them voting 'No' to the power plant, the opposition movement began to gain strong momentum— after 25 years of 'defeats' and with only two supporting members of the council versus twenty opponents. The independent referendum was followed by a tremendous victory in the town council elections held in April, with 12 out of 22 elected assembly members opposing the proposed nuclear plant and supporting an official referendum. When it became public that Mayor Satö had attempted to rush through the land-sale after the independent referendum, the new council commenced moves to officially recall him from office. He resigned on 15 December 1995, just seven days after the Monju incident, which questioned the security of nuclear facilities, and before the recall signature verification process was completed. The tide had turned. Residents who had reservations about the nuclear power plant were gradually provided with political opportunities to express their concerns, and the opposition movement successfully seized the majority of the residents.

An explanation of the structure of political opportunities

As Doug McAdam notes (1996:27), the structures of political opportunity can be interpreted in many ways, but one of the more useful is an analysis along four dimensions:
  1. the relative openness or closure of the institutionalized political system;
  2. the stability or instability of that broad set of elite alignments that typically undergird a polity;
  3. the presence or absence of elite allies that supports the challengers; and
  4. the state's capacity and propensity for repression.
This framework is usually applied to the national political level—for example, by Herbert Kitschelt (Kitschelt 1986), Hanspeter Kriesi

(Kriesi et al. 1995) and Sidney Tarrow (Tarrow 1994), but here I apply it to an analysis of the regional and local government levels.

Relative openness
A unique factor in the Maki Machi dispute was the presence of the town owned land at the center of the proposed reactor site. The Local Autonomy Law regulates the sale of municipally owned property, and states that the sale of any town property larger than $5,000m^2$ requires the approval of both the council and the mayor. This piece of municipality owned land thus offered the Maki Machi residents a unique opening to resist the developments from within the political system.[4] Many of the events that unfolded in Maki Machi were determined by phenomena at the institutional level. Single term mayors had come and gone, the safety assessment process had been stalled since 1983 and the construction project was therefore on hold, a new council moved to recall Satö and the necessary ordinance to hold the referendum was passed. Perhaps the most astonishing impact of the unofficial referendum was its opening of the stultifying political system through the election of new councilors. As Yüko Takubo observed, the structure of political opportunities is not fixed, but 'can be altered and created anew by social movements' (Takubo 1997: 139).

The executive's capacity to implement
A government's capacity to implement their program is determined to some extent by the factions and conflicts among the power elites. As has been frequently observed, two conservative factions had dominated politics in Maki Machi since the early 20th century. Rival factions formed around two long-standing conservative members of the national Diet, each with a solid grip on regional power. Each faction included members of the prefectural assembly, local town councilors, and construction companies. Their respective strength was evenly matched, making the political situation in the area extremely volatile, especially around the time that the plans for the nuclear power plant were announced. When Satö was reelected in 1990 he became the first mayor in sixteen years to serve a second term. At every election since 1974 the mayoral candidate from the rival faction, fearful that the others might reap the benefits of constructing the power plant, had proclaimed a moderate and cautious stance on the issue and defeated the incumbent. Some 2,000 anti-nuclear 'swinging' voters who were independent from both

conservative factions held the deciding votes. After each of these elections, the new mayor would soon begin to push the pro-nuclear case. The other faction would then seduce the anti-nuclear camp with a more moderate position and win the next election. Thus, until Satö's reelection in 1990, the project remained at a standstill, as the two conservative factions fought among themselves with increasingly moderate candidates. In other words, for many years the greatest obstacle to the Maki No. 1 reactor project was the political infighting among the local power elite and the resulting instability.

The reshuffle in national politics and the subsequent volatility of town politics

Less frequently noted is the increased political volatility in Maki Machi following a reshuffle in national politics. Niigata Prefecture was the home of the former Prime Minister Kakuei Tanaka (1918–1993) and had long been the stronghold of the LDP's Tanaka faction until Tanaka suffered a brain infarction in 1985.[5] After Tanaka was immobilized, the prefecture lost its centripetal power base.

There was a split in the ruling Liberal Democratic Party in June 1993, and in October 1996 the electoral system was reformed and electoral boundaries were redrawn. Tatsuo Ozawa, a former Minister for Health and Welfare and the head of one of the Maki Machi factions, defected from the LDP in June 1993 with fellow former Tanaka faction members including Ichirö Ozawa, a powerful leader. Tatsuo Ozawa joined the new Shinseitö Party and then the Shinshintö Party, but soon parted company with Ichirö Ozawa and others to form a party named the Kaikaku Club. Reforms to the electoral system about this time introduced a single seat constituency system, which involved redrawing the electoral boundaries. The Niigata electorates were redrawn such that Maki Machi moved from the old No. 1 electorate to the new No. 2 electorate, which also included parts of the former No. 3 electorate. After these reforms of October 1996, Tatsuo Ozawa left his electorate and won a proportional representation seat in the Lower House. Shin Sakurai, the former LDP member for the Niigata No. 3 electorate, won the Lower House seat for the new No. 2 electorate. Many local and regional politicians in Maki Machi switched their factional allegiances from Ozawa to Sakurai around this time.

In the midst of all of this, in February 1994 Motoji Kondö, a former Minister for Agriculture, a leading member of the Miyazawa faction and the head of the other conservative faction in Maki Machi, passed away. This political turmoil between 1993 and 1996 at the national level virtually coincided with the collapse of local power structures in Maki Machi and Niigata Prefecture, as the area's two rival factions were dismantled and reorganized. The sitting governor in 1992 had been forced to resign in the wake of bribery allegations. In August 1996, the incumbent Governor Masao Hirayama's position was shaky as he campaigned for a second term in office. By August 1996, when Maki Machi residents voted in the local referendum, Ryutarö Hashimoto of the LDP had been in power as prime minister for seven months in a coalition with the Socialists and the Sakigake Party. The LDP came back into power in July 1994, after its first loss of office in almost 50 years, in a coalition headed by the Socialist leader Tomiichi Murayama as Prime Minister. In Maki Machi, the grip on power by the two ancient factions weakened, and the residents' campaign for a referendum arose in this opening, gathering momentum and eventually succeeding.

Elite support for challengers
The 'Group to bring about a local referendum' arose as a new movement initiated by independent business owners and local leaders who were not far removed from the core political power structures in the town. The main opposition movement had been initiated by the 'usual suspects'—the labor unions and the political left. Before the mayoral election of August 1994, the movement was well-aware of the problem of their limited support base, but had little success in expanding it. The new 'Group to bring about a local referendum', however, succeeded in drawing support for the anti-Maki nuclear reactor cause from the town's more conservative residents. I will return to this when discussing the cultural framing of the movement.

Social control
The existing power factions at all levels of government in Maki Machi—national, prefectural and local—were in extreme states of dynamic flux in the mid 1990s. Thus the residents and the town were relatively free of control from external ('higher') authorities, and could take risks without fear of reprisals. Increasing demands for the decentralization of power—i.e. giving more power to regional authorities—also helped to alleviate fears and anxieties about central government interventions.[6]

Maki Machi's mobilizing structures

In resource mobilization theory, the term 'mobilizing structures' refers to a group of variables that define a social movement. In this section I will analyze the resources available to the Maki Machi anti-nuclear movement.

The movement opposing the nuclear power plant at Maki Machi had abundant human, network and information resources. Typically, as in the Rokkasho Mura case (to be discussed shortly), nuclear power plants and facilities are located in remote and depopulated regions. Opposition to power plant proposals in these highly marginalized locales is largely dependent upon local landowners, farmers and fishermen. The human and information resources available to be mobilized are generally quite limited. Maki Machi, however, is one of the rare exceptions to this rule in Japan. As the county seat of Nishi Kanbara Gun, and with its close proximity to the prefectural capital, Maki Machi has long been prosperous. Substantial resources were therefore available to the opposition movement.

As a regional center, Maki Machi has four public high schools, three sake breweries and the branch office of a regional newspaper. More recently it has become a 'satellite-dormitory suburb' for the city of Niigata, especially since the openings of the Jōetsu bullet train line and the Kan'etsu and Hokuriku freeways. These local peculiarities provided an important backdrop to the mobilization of public opinion against the proposed nuclear power plant.

The primary actors in the 'Group to bring about the local referendum' were local self-employed business leaders, largely from the first baby boomer generation. Local high school teachers, lawyers, doctors, dentists, university lecturers and other professionals, generally of the same generation, played pivotal roles in the movement. This is significant, because in other areas where there has been opposition to nuclear facilities or other public works, the professionals involved in the movements are generally 'outsiders' who reside in a capital city of the prefecture or distant metropolitan areas. At Maki Machi, however, these participants were, for the most part, born and raised locally.

Its proximity to the prefectural capital, with its population of half a million and a national (Niigata) university, helped to keep the dispute in the public eye. For example, Maki Machi residents were able to enlist the help of Professor Akane Akita, then with the law department at Niigata University, in drafting the local referendum ordinance. Many self employed professionals living and working in Niigata City and nearby, including a group of doctors and dentists, also contributed significantly to the Maki Machi movement.

Maki Machi's population is slightly more than 30,000, but from the perspective of economic resources, its residents' resistance movement benefited from its very small size. According to one of the movement's leaders, inserting leaflets in local home delivered newspapers to every household in town was feasible because of the town's size. Delivering a leaflet by inserting it in the newspaper to all 8,700 households only cost about 50,000 yen. The size of the town has allowed the movement to survive and continue.

Framing: Self-determination and grassroots alternatives

'Frames' refer to commonly shared 'definitions of a situation', a 'worldview' or the 'self-image' of an individual, a community, a populace, or a social movement. As discussed in chapter 5, there must be some correspondence between an individual's cultural frames and the frames of the movement if the former is to be motivated to join in the latter's collective action. Snow refers to the conscious and strategic formulation of self-images as the process of framing (Snow et al. 1986). One of the difficulties that anti-nuclear power movements have in attracting wider support from the community can be attributed to the problem of creating new and 'positive' frames to counter their inherently oppositional and critical images—to demonstrate that they represent something more than simply opposing or rejecting 'development' and 'progress'.

Maki Machi's anti-nuclear movement entered a new stage in the lead up to the mayoral election in March 1994 with an 'origami crane campaign' initiated by the town's women. By May 80,000 origami cranes had been given to Mayor Satō as signifiers of the women's concerns. These traditional symbols of peace framed the opposition to the horrifying risks of nuclear power in terms of ordinary citizens, especially women, who were seeking alternatives to the society being created by the pro-nuclear government-industry alliance. At the same time, folding a piece of paper into an origami crane was something that anyone could do, and through which they could anonymously express their opposition to the proposed nuclear power plant. It was also relatively easy to persuade friends and family members to contribute. The 'origami crane campaign' preceded the August 1994 mayoral election, and eventually culminated in a remarkable achievement in April 1995 when three women anti-nuclear candidates topped the local council elections. The 'origami crane campaign' proved to be an ideal frame for attracting women to join the movement. One of the candidates in the 1994 mayoral election, Mr. A., had no prior political involvement. Nor was he particularly active in the anti-nuclear movement. He organized the 'Blue sea and green group' to support his campaign, the name signifying appreciation of Maki Machi's natural beauty. Playing the guitar and singing songs written by his comrades during his campaign, Mr. A. proved to be quite popular among the younger and newer residents, polling an unexpectedly high 4,382 votes. Their slogan was very direct: 'No nukes'.

The push for a 'regional referendum' also became a new frame for the movement. As previously mentioned, the town assembly was dominated by proponents of the power plant, and the chances of getting the necessary ordinance passed to hold the referendum appeared to be quite slim. The 'Group to bring about a local referendum' therefore staged an independent referendum, mimicking an official one in every possible respect, except that the voting was over a two week period between 22 January and 5 February 1995, rather than one day. The voter turnout was much greater than expected—10,378 people, 45.4% of the eligible voters participated. Of those, 9,854 (95%) voted 'No' to the nuclear power plant. This huge turnout and the overwhelming majority of the 'No' votes gave the movement momentum. When the incumbent Satō polled only 9,006 votes to win the 1994 mayoral election, there was reason to be optimistic that he could be defeated the next time.

The referendum allowed people to express their opposition anonymously, and yet made the size of the opposition movement apparent for all to see. The 'Group to bring about a local referendum' had taken care to differentiate themselves from the older anti-nuclear movement and avoided frames such as 'opposition to the nuclear power plant' or 'the campaign to halt construction'. Instead, adopting a set of scales for the group's logo, they concentrated on maintaining an image of neutrality while facilitating the official referendum, in order to maximize voter participation. They did not reveal the group's true position on the plant until the referendum was finalized. Their neutral image effectively avoided tainting the campaign with old and tired political colors, and helped to ensure the referendum's legitimacy as a democratic means of enabling people to determine their futures for themselves. Although the referendum was not legally binding, its advocates were well aware that its political influence was highly dependent upon the voter participation rate. As we have seen, their strategic cultural framing proved very effective, with a much larger voter turnout in the official referendum (88.3%) than expected. In the wake of the unofficial referendum, another frame emerged in which Mayor Satö and Töhoku Electric were portrayed as 'heavy handed' in their refusal to sanction an official referendum. They dismissed the residents' call for an official referendum as legally groundless. Satö's response to the unofficial referendum was to try to rush the land sale, calling an extraordinary town council meeting on 20 February 1995 to seek approval. It was prevented only by a one-and-a-half-day hunger strike and other actions by local residents, including Mrs B., a leader in the 'origami crane campaign'. After the council finally passed the referendum ordinance, Satö masterminded an amendment that changed the timing of the referendum from 'within 90 days of the bill's passage' to 'such time as both the mayor and the assembly agree upon'. On 16 October, when pressed by residents about exactly when it would be held, Satö responded, 'I have not given it any thought. The sale of the town land has nothing to do with the referendum'. Satö's heavy handedness, and the probability that the proponents of the power plant were planning to scrap the referendum bill and approve the land sale in the regular council meeting scheduled for December were significant factors leading to the moves to recall Satö from office (between October and December 1995).

The great earthquake that devastated Kobe only days before the unofficial referendum in January, and the accident that shut down the Monju fast breeder reactor only days after the signatures for Satö's recall were submitted on 8 December both provided potent frames, casting fundamental doubts about the safety and reliability of mammoth technologies such as nuclear power.

Another potent framing activity began two months before the official referendum in June 1996. Mrs. B. and others began a new campaign, called the 'tree of happiness'. Handkerchiefs with anti-nuclear messages were tied to ropes and hung from the tops of posts. People freely wrote their thoughts about the nuclear power plant on the handkerchiefs. This campaign clearly demonstrated the growing support for the opposition movement both within and outside of the town, as the first 'trees of happiness' in town continued to grow with more ropes and more handkerchiefs, and additional 'trees of happiness' were created around the community. Handkerchiefs thus provided a frame representing grassroots opposition to technology and the nuclear power plant.

The difficulties at Rokkasho

The marginality of Rokkasho

Since 1989 I have been involved in studies of the changes that have occurred in Rokkasho Mura, focusing on the progress and social impact of the Mutsu Ogawara development project[7] and the conflict arising from opposition to the nuclear fuel processing facilities (see Figures 9.2 and 9.3). Comparing Rokkasho Mura's predicament with Maki Machi, the difficulties facing the former are painfully obvious. Maki Machi is a regional center, closely neighboring a mid-sized prefectural capital city to which it is well connected by public infrastructure and social networks. It is therefore only slightly 'peripheral'. In contrast, Rokkasho Mura is clearly marginalized geographically, socially, and historically. It is geographically rugged and remote, isolated from any large population centers. There was no senior high school, no sake brewery, and no newspaper branch office in the village until the Mutsu Ogawara development project began (a senior high school was established after the Mutsu Ogawara development but even today there is no sake brewery, and no newspaper branch office in the village). Most of the first baby boomers born there left the village, and those who remain are either employed by the village hall or are either directly or indirectly involved with nuclear fuel processing or related facilities. Very few can freely express their opinions, due to local politics or economic dependency on the development project.

Disadvantaged by their remote location, only a small proportion of students continued to secondary or higher education. Very few are professionals. Within this village, the main participants in the movements that opposed the Mutsu Ogawara development and the nuclear facilities were largely farmers and fishermen. Once the land for the nuclear facilities had been acquired and compensation for the fishermen completed, the possibility of any new political or procedural opportunity for obstructing or halting the project was quite limited. To put it simply, the movement opposed to the Mutsu Ogawara development and the nuclear facilities faced enormous difficulties (see Hasegawa 1998a).

Closed political opportunities
With such limited possibilities within the official channels, the
opportunity to stop a coastal study in the summer of 1986 provided a
major turning point for the anti-nuclear movement in Rokkasho Mura.
The coastal study was part of the environmental assessment required
before construction could begin, but it could not be conducted
without the approval of the fishing co-operatives. Fierce conflicts
broke out in the Tomari fishing co-op, the biggest of the three co-
ops in the village, over whether to permit the study or not. Both the
fishing co-op and the Tomari community itself were deeply divided
over this issue. Fishermen were arrested for attempting to stop the
study, the chief of the co-op was dismissed, and chaos swept through
this community. Once the coastal study was completed, the number

Figure 9.2   Shimokita Peninsula and nuclear facilities

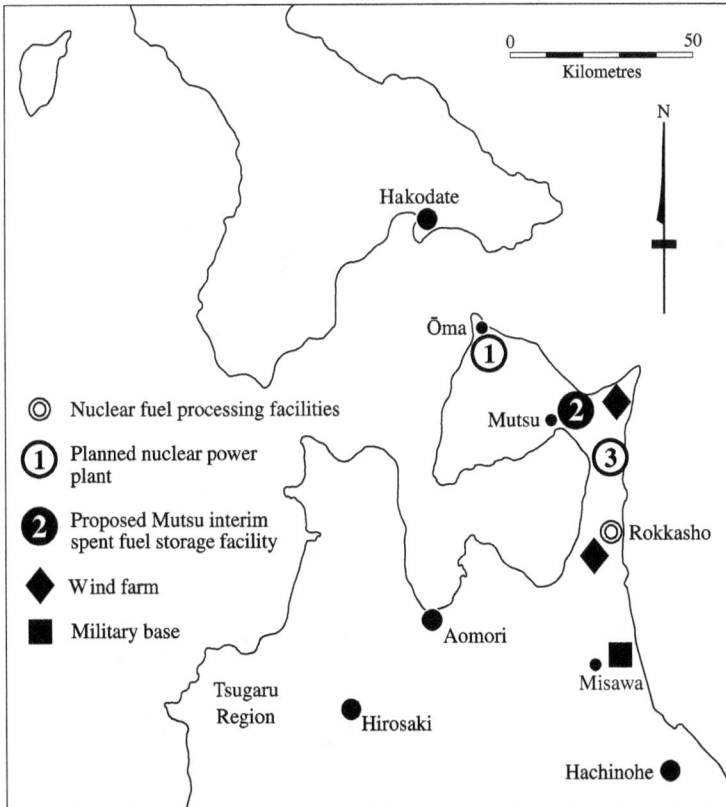

Figure 9.3   Rokkasho Mura and nuclear facilities

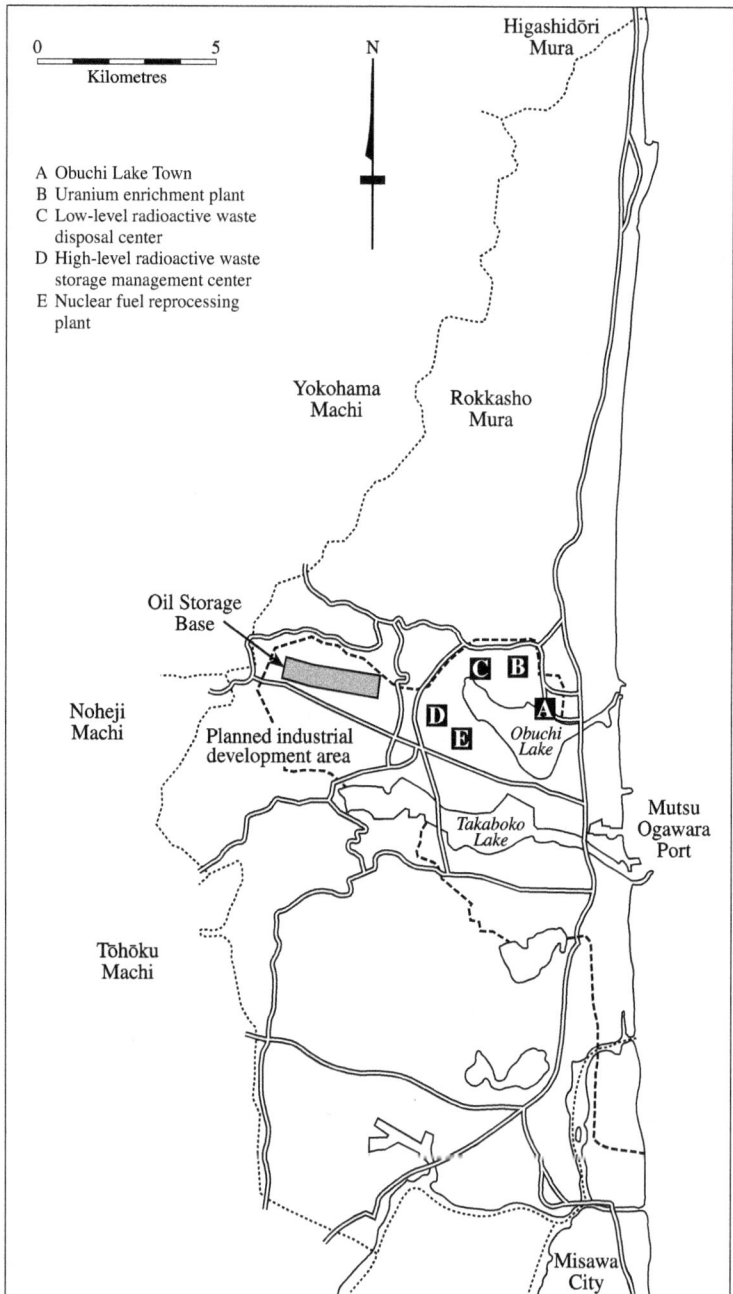

0   5   N   Higashidōri
                                              Mura
Kilometres

A Obuchi Lake Town
B Uranium enrichment plant
C Low-level radioactive waste
   disposal center
D High-level radioactive waste
   storage management center
E Nuclear fuel reprocessing
   plant

Yokohama   Rokkasho
Machi   Mura

Oil Storage
Base   C   B

Noheji   D   A
Machi   E   Obuchi
        Planned industrial   Lake
        development area

                          Takaboko   Mutsu
                          Lake   Ogawara
                                        Port

Tōhōku
Machi

                              Misawa
                              City

of local residents still remaining in the opposition movement was getting smaller and smaller.

The proposed nuclear fuel processing facilities included a spent fuel reprocessing plant, a uranium enrichment plant and a low level radioactive waste storage facility. Table 9.1 provides an outline of the project. The legal basis for proceeding with these facilities was an agreement signed in April 1985 between the Governor of Aomori, the Mayor of Rokkasho Mura, and the three operators' companies.

Table 9. 1 Outline of the nuclear fuel processing facilities[a] in Rokkasho Mura

Type of facility	Capacity	Operations commence	Construction Cost	Estimated workforce
Nuclear fuel reprocessing plant[b]				
Main plant	800t pa	(2006[c])	≈2,140 billion yen[d]	
Spent fuel storage pool[e]	3000t	1997		≈2,000 (in operation)
High-level radioactive waste storage management center	1440 canisters (vitrified high-level waste) 2880 canisters (future)	1995	80 billion yen (for 1440 canisters)	
Uranium enrichment plant[f]	150t (initially) 1050t pa (current)	1992	≈250 billion yen (for 1500t pa projected peak level)	≈1,000 (construction) ≈300 (operations)
Low-level radioactive waste disposal center[g]	≈1 million drums (3 million drums in future)	1992	≈160 billion yen (for 1 million drums)	≈700 (construction) ≈200 (operations)

Notes:
a Table compiled from the Japan Nuclear Fuel Service Co.'s data; current at the end of December 2003.
b After nuclear fuel is used in a reactor for four years, it becomes what is known as 'spent fuel'. The Japanese government plans for this spent fuel to be sent to reprocessing plants where the reusable uranium and plutonium will be extracted and separated from the unusable nuclear fission products, which are called 'high-level radioactive waste'.
c The date for commencing commercial operations, first projected for December 1997, was postponed six times by the end of 2003. About four kiloliters of water may have leaked from a spent fuel storage pool over a six month period beginning in July 2001. Subsequently, 291 instances of substandard welding were identified, including the one responsible for the water leak. Furthermore, there has been a review of 500,000 instances of suspected corner-cutting. Due to these accumulated problems, although officially the commencement date is set for July 2006, at the time of writing it is highly likely that it will be postponed for at least a further twelve months.
d The delays in commencing operations have been accompanied by budget blow-outs, with the estimated costs increasing many times. The cost in 1999 was estimated to be 2,140 billion yen. The most current estimate is in excess of 3,000 billion yen, about four times the original estimate of 840 billion yen.
e The pool is for the temporary storage and cooling of spent nuclear fuel before reprocessing.
f Only about 0.7% of natural uranium is the so-called 'burnable' uranium isotope, U-235. In order to use it as fuel in a nuclear power plant, enrichment is necessary to increase the proportion of U-235 to between 3 and 5%.
g The function of the center is to bury and manage low-level radioactive waste that is produced during nuclear power generation. The plan is to manage the waste at Rokkasho for 300 years.

The unexpected national House of Councilors election result in July 1989 seemed to indicate a change in the political tide, raising hopes that the incumbent mayor and governor could be ousted, and the agreement scrapped. The mayoral election in December 1989 and the gubernatorial election in February 1991 attracted a lot of attention. Various opinion polls indicated that the majority of the local population was either definitely opposed to the facilities or had a negative view of them. Hiroshi Tsuchida won the mayoral election after promising to 'freeze' the project, but once in office, explained that 'to freeze' meant to move forward slowly and steadily. He then signed-off on the safety agreement required for the construction of the facilities. The incumbent Governor Masaya Kitamura won a fourth term in the February 1991 election, bringing the construction of the nuclear facilities that much closer to being a *fait accompli*. The opposition movement lost all but one subsequent election until 2003. The exception was in 1993 when an anti-nuclear candidate won a seat in the national Lower House election. Most noticeable was the disappearance of votes for the Socialist Party, the principal supporter of movements opposing the Mutsu Ogawara development project and the nuclear fuel processing facilities. The Socialists stood 13 official candidates and supported two independents in the 1991 Prefectural Assembly election. They suffered a disastrous result, with only one official candidate and one supported candidate winning seats, each in unchallenged electorates. The Communist Party lost seats in the same election, drastically reducing the number of assembly members opposed to the nuclear fuel processing facilities from 11 to only three, including one independent. These three seats have been retained by the opposition movement candidates until the present.

Simply in order to continue to appeal to people who oppose the project, a social movement should demonstrate the strength of its support base by fielding its own candidates for elected office. But following the creation of a new national peak labor organization, Rengö, in November 1989, and numerous changes in the national political landscape as discussed above, the Socialist Party has lost much of its support. This has led to a vicious cycle—with no prospect of winning, election campaigns tend to be very low key operations, thereby attracting ever diminishing participation, resulting in even fewer votes than expected, which in turn saps more energy from the movement.

Governmental capacity of the executive

At the other end of the political spectrum, when we analyze Rokkasho's social movement in terms of the tensions and conflicts among the ruling elite, there were schisms in the conservative camp at all levels of politics in the region around 1989–91. An anti-nuclear candidate managed to poll an unprecedented 370,000 votes in the House of Councilors election in July 1989, thanks in part to such a schism. But in every other election the opposition movements have failed to capitalize on the conservatives' disunity. The Rokkasho mayoral election in 1989 saw a conservative incumbent unsuccessfully defending himself against another conservative candidate, Tsuchida. The conservative camp was divided in the February 1991 gubernatorial election, and again failed to find a candidate acceptable to both sides in February 1995, when Morio Kimura defeated the incumbent Governor Kitamura. Mayor Tsuchida was beaten by the conservative Hisashi Hashimoto in December 1997. The anti-nuclear movement here has yet to capture the conservative vote, in contrast to Maki's movement after 1995.

Elite support for challengers

The only notable support for the Rokkasho opposition movement amongst the local power elites came from Nökyö, a powerful agricultural co-op. Around 1987 and 1989 when the opposition movement began to gather momentum, the leaders of the prefectural branches of Nökyö supported the anti-nuclear movement. This provided a significant support base, including most of the local farming community, and especially the youth and women's section of Nökyö, which were very strong and active in the opposition movement. But it also had significant limitations as it subjugated individual farming families' opposition to the very formal Nökyö structures, especially its decision making processes and broader political objectives. Thus unlike a typical residents' or citizens' movement, there was no spontaneous eruption of energy from this sector. Mobilizing these particular resources was entirely dependent on the political leadership and persuasions of the Nökyö leaders. Over-reliance on this source of support, then, meant that the movement risked being stifled whenever the Nökyö leaders decided that their rank and file members needed to be put back in their places.

Social control

The opposition movement in Rokkasho had been confronted with a pro-development village administration during the four terms of Mayor Isematsu Furukawa (1973–89). Furukawa's administrative style was dictatorial, using both overt and covert methods as required to overcome any resistance to his political agenda. (For example, on the pretext of the threat to job opportunities, Furukawa's administration refused to allow opposition groups to use public halls and facilities.) Rokkasho Mura covers a vast area but its population lives close together in hamlets dotted around the region. In close knit communities such as this, people tend to be highly aware of each other's activities and there is little room for anonymous action. Voicing one's opposition to either the Mutsu Ogawara development or the nuclear facilities could lead to social isolation and family members losing their jobs. Both projects had been approved at the national level and were promoted as 'national projects', making them more difficult to oppose in Rokkasho Mura and Aomori Prefecture. At both the local and prefectural levels of government the dominant political framework was nationalist and pro-development—and hence any expression of opposition to these projects was treated as against the public interest or "un-Japanese". These frames exert strong pressure on citizens to suppress their non-compliant thoughts and feelings.

The national government and the LDP politicians took advantage of the governor's subservient attitude (whether conscious or unconscious) to push through the nuclear fuel processing facilities, the Mutsu Ogawara development plan and to extend the bullet train line to Aomori City with the frames of 'national interest' and 'economic development' effectively suppressing almost all dissent.

Mobilizing structures: Supporters and networks

Rokkasho Mura's marginalized anti-nuclear movement had few resources, but they were supported by the national anti-nuclear movement through organizations such as the Citizen's Nuclear Information Center in Tokyo, and the residents of cities with populations of 2–300,000, such as Hachinohe, Aomori and Hirosaki in Aomori prefecture. Here as elsewhere, numerous 'outside' professionals, especially university academics and lawyers, played pivotal roles, relatively free of the constraints of local political, economic and family ties.

The 'Information Center on the issues of the nuclear fuel processing facilities for the residents of Aomori Prefecture' was established in February 1987 to closely and critically monitor the activities of the prefectural government and the facility operators, reporting developments of the project to the opposition movement in a bimonthly newsletter. The Center and newsletter continue to this day. An administrative lawsuit initiated against the four nuclear facilities, by lawyers and protestors in Hachinohe and surrounding areas of Rokkasho, and with the 'Group of 10,000 supporters of legal suits to stop the nuclear fuel cycle facilities', has become one of the biggest legal battles ever fought over the Japanese government's nuclear policy.

When the opposition movement was at its peak, marked by its positive results in the Upper House elections in 1989, it involved an alliance of various groups—a citizens' movement of urban professionals plus labor unions and farmers. The farmers' opposition always remained independent, though, keeping its distance from the citizens' movement. Only a handful of their leaders had any contact with the citizens' movement, the political parties, or the unions. The farmers' opposition was also largely limited to one region of the prefecture (Tsugaru), with very few members in Rokkasho Mura or its surrounding areas. And even this opposition waned, especially after 1991 when farmers in the Tsugaru region were devastated by typhoon No. 19 and reconstruction (via government subsidies) became their top priority.

As elsewhere, women provided the foundation for the mobilization of human resources (Iijima 1998a). At the peak of the campaign to halt the coastal study in 1986, inspired by a post-Chernobyl sense of urgency, the wives and mothers of fishermen in Tomari hamlet—known as the 'Moms' Corps' (Kattcha Gundan)—made their presence felt through protest activities such as collecting signatures, sit-ins and choral speaking.

From 1987 to 1989, the women's and youth sections of Nōkyō played central roles in the opposition movement. This was the peak of the movement in Aomori Prefecture. Two women leaders from the opposition movement stood as candidates, one each for the Socialists and Communists in the 1995 gubernatorial election. The 'Mothers group protecting children from radiation' has held monthly demonstrations in Hirosaki City for the past 18 years, beginning with the 'Women's demonstration to oppose the nuclear fuel cycle facilities and nuclear power plants' in July 1986. The group placed a full-page advertisement in four regional newspapers in April 1987 and collected some 2.8 million yen from 1,500 donors across the country. Two other from Hirosaki City placed an advertisement in the *Washington Post* to express their opposition to air freighting plutonium into Misawa Airport. They also produce a newsletter from time to time. The nationwide growth in anti-nuclear sentiments following Chernobyl was predominantly among women, and the women of Aomori did their bit to oppose the nuclear facilities proposed for their prefecture (see chapter 8). Since 1991, the anti-nuclear movement has receded from this national peak, but the women in Aomori are still fighting at the time of writing in 2003.

Beyond negative framing

When Rokkasho Mura is depicted on television, the pictures are typically of vast areas of the vacant land where the failed Mutsu Ogawara development project might have stood, or of the nuclear facilities under construction, or of suspended operations due to industrial unrest or accidents. The village is frequently referred to as a 'nuclear waste dump'. Visual documentaries about the village tend to open with images such as 'an old lady with a bent back braving the blizzard alone'. They invariably describe the *yamase*, the cold gales that sweep the region in summer. Symbols with negative connotations are emphasized, such as depopulation, the *yamase*, the cold summers, poverty, the residents' need to work away from home, nouveau riches from property sales, development refugees, flight by night, and abandoned houses. In short, Rokkasho Mura is typically framed in the media by its marginal and tragic characteristics.

The opposition movement also emphasizes Rokkasho Mura's tragedies, attempting to demonstrate the urgency of the current situation to the broader public through the mass media. But this kind of negative framing makes it very difficult to win the support of the wider community, who tend to see it as simple negativism or opposition to change that does not offer any positive alternatives for the future. Maki Machi was able to present a positive alternative framed by the desire to preserve the natural beauty of the area. In contrast, with the center of the village presently occupied by a vast wasteland, that argument is not very persuasive in Rokkasho Mura.

Their proponents claim that the Rokkasho Mura nuclear facilities are cutting edge technology, but simple technology worship is no longer sufficient to win the hearts and minds of the affected population. The opposition movement in Rokkasho Mura now face a *fait accompli*—all of the proposed nuclear facilities except for the reprocessing plant, which is due to commence operations in July 2006 after frequent deferrals, are already built and in operation. The remaining opposition must now confront the huge task of creating a viable alternative future for Rokkasho Mura that does not include nuclear facilities.

10 The dynamics of social movements and official policies: Green electricity

Defining green electricity

Having been involved as an advisor to an NGO promoting green power schemes since 1996, in this chapter, I will examine the dynamics of this sector of the environmental movement and related government and industry policies based on my own experiences and observations.

The concept of 'green electricity' has been attracting worldwide attention since the mid 1990s. It is an important part of the green consumers' movement, wherein consumers willingly pay higher prices for environmentally friendlier products. Consumer demand sends clear messages to producers, distributors, retailers, the market and governments that there is a market for such products. Green consumers' movements also support the production and sale of organic agricultural goods, recycled paper and other recycled goods, while various NGOs and government agencies certify products that meet minimum environmental standards by issuing 'environmentally friendly' labels.

Until quite recently electricity was not considered to be the type of product for which consumers had any choice. Electricity was generated and distributed mainly by private or public utilities with regional monopolies. Consumers could thus either sign-up with their local suppliers or do without electricity. Once signed-up, consumer choice was limited to controlling their consumption or using reduced price off-peak power. Under the system of regional monopolies, ordinary citizens had little opportunity to affect the decision-making processes of the utilities unless they became activist shareholders who asked awkward questions at shareholders' meetings.

Electricity is an essential foundation of modern lifestyles and its supply is therefore clearly a 'public good'. But at the same time, generating electricity is one of the most environmentally damaging industries in modern societies. Constructing huge hydroelectric plants invariably entails the forcible relocation of people from their ancestral homes and communities, radically alters the landscape, destroying forests and the ecosystem for many plants and animals. Thermal electric plants are among the world's leading sources of air pollution, discharging massive amounts of carbon dioxide and a cocktail of other toxic chemicals into the atmosphere. Nuclear plants entail the risk of low level radioactive pollution even when operating normally, and a serious accident could have catastrophic affects. Disposing of radioactive waste material is also a significant problem. Since the early 1970s, nuclear power has been one of the most contentious issues in Japanese society, as in Germany, the US and other developed countries. In recent years, the power utilities' annual shareholders' meetings have taken longer than any other industry's. But objections come only from minority shareholders with little real power to affect the company's policy direction.

So what can consumers do to make the utilities and the market aware of their desires for a better service? One option is to purchase green electricity. Green electricity is the electricity generated from renewable power sources such as wind and solar power. As a framing device, calling it 'green' differentiates renewable energy resources from non-renewable and other environmentally hazardous sources such as nuclear power, oil or coal fired thermal power and large-scale hydroelectric power. The term 'renewable energy' is not a very catchy slogan—too cumbersome and technical to capture the ordinary consumers' imagination. 'Natural energy' is commonly used in Japan but is not very common in English. 'Clean electricity' has become a worn out cliché through its use to promote the nuclear power industry and is therefore no longer credible. 'Green electricity' directly engages the public's imagination, evoking cleanliness and naturalness without the negative connotations that have become attached to the terms clean and natural. Green is an internationally recognized symbol for environmental protection and ecological values. It is used in various fields and in various ways to invoke environmentalist values, such as 'greening the auto tax', 'greening America', 'green politics', 'green Christianity' and so on.

An important difference between many 'green' goods and green electricity is that for the latter there is absolutely no difference in the physical characteristics of the products. Electricity from renewable sources and from more conventional thermal and nuclear plants comes down the wire together. Hence for the consumer 'green power' is purely a notional differentiation.

The great advantages of renewable energy resources are that they do not emit carbon dioxide, do not produce radioactive contaminants or waste and do not run out. For a long time, though, the cost of generating power from them was prohibitive.[1] Thus the greatest obstacle to more wide spread development of renewable energy technology has been the question of who will bear the relatively higher cost of power. Various green electricity schemes have been developed to overcome the associated problem. Since there is as yet no well-established definition, I offer the following: a 'green power scheme' is a social system that directly conveys the costs of generating electricity from renewable resources to the consumer or taxpayer (on a voluntary or 'opt-in' basis).[2] Various green power schemes are currently being developed and trialed in a number of countries around the world, especially in developed countries.

The 'PV (photovoltaic) Pioneers' program established by the Sacramento Municipal Utility District (SMUD) in California is widely recognized as the world's first scheme of this type. The scheme was initiated by SMUD's management in an attempt to secure an alternative power source and win back the trust of the local residents—both necessary to rebuild the utility after a 'citizens initiated referendum' had forced it to shut down its nuclear power plants in June 1989. The south facing roofs of the willing participants' houses were fitted with solar panels that were connected to the grid. The electricity generated by the panels is not directly consumed by the particular household but is distributed through the grid. In exchange for their roof space and a monthly fee of four dollars, the participants receive information about ways to reduce power usage and solar power in general. They also receive special benefits, such as priority treatment when installing solar water heaters and public recognition as 'PV Pioneers', but they do not receive any direct financial benefit, such as reduced electricity rates or credit for the power generated. From an economists' perspective, these people are 'good natured to a fault'. Nevertheless, the scheme spread the expense of generating solar power across the interested consumers and the utility. Consumers participated in the scheme to support the radical new non-nuclear direction taken by SMUD, which included using more renewable energy resources while encouraging power saving through the more efficient use of electricity (Hasegawa 1996c: 149–56).

The beginning of the green power movement in Japan

The Seikatsu Club Hokkaidō consumers' co-op

Japan's first green electricity scheme was introduced by the Seikatsu

Club Hokkaidö consumers' co-op. The Seikatsu Club consumers' co-op was established in 1965 in the Setagaya district of Tokyo. Its initial aim was to purchase good quality safe milk. Since then the co-op has promoted many safe and healthy agricultural products, processed foods and other consumer goods.[3] Yoshiyuki Satö and others have conducted extensive sociological analyses of the Seikatsu Club co-ops in the Tokyo metropolitan area (Satö ed. 1988).

The Seikatsu Club Hokkaidö co-op was established in 1982 and currently has 13,000 members. It is one of the most active affiliates of the Seikatsu Club in terms of energy issues in general and nuclear power in particular. This particular focus of their activism was triggered in 1987 by the discovery of radiation in their pesticide-free tea products, presumably caused by the fallout from the Chernobyl accident the previous year. This discovery ignited a prefecture wide campaign in 1988–89, demanding a referendum over the commencement of the operation of the nearly finished Nos 1 & 2 nuclear reactors at the Tomari power station. The campaign was primarily conducted by the co-op's female members and the local trade unions who circulated petitions, aiming to collect one million signatures. Although they surpassed this target by 30,000, the 110 seat Prefectural Assembly nevertheless rejected the referendum ordinance by two votes. Since 1990, about 50 co-op members have taken part in a summer camp each year in Horonobe Chö, where it was feared that a high-level radioactive waste dump site was to be built. Their mission is to go door knocking in the town and the surrounding area to inform the local people about potential safety problems and other negative impacts of the project. The Tomari power station and Horonobe facility are typical of the disputes about nuclear energy in Hokkaidö. After the narrow defeat of the referendum ordinance, the co-op members decided that they needed their own representatives in local councils. Since winning its first three seats in the Sapporo city council and one in the Ishikari Chö town assembly in 1991, it has supported numerous successful candidates for local assemblies through a sister organization, the Seikatsusha Network (The network of residents).

When it became apparent in July 1996 that Hokkaidö Electric was planning to build a third reactor at the Tomari station, the female members of the co-op began an opposition campaign—a new petition, this time seeking only 100,000 signatures. During this part of the campaign, I was invited to speak at a public meeting on 15 October 1996. My discussion was entitled 'What citizens can do to foster renewable energy with an extra 10%: An introduction to the green power scheme'. I began by outlining the green power scheme

introduced by SMUD, and then discussed various problems with
Japan's nuclear power policies. I suggested that a new consumer
movement should be launched, in which citizens would pay a 10%
surcharge on their electricity bills towards a fund to foster renewable
energy. The surcharge, I suggested, could be offset by the consumers'
efforts to reduce their personal/household energy consumption,
and hence their power bills. 'Beginning with milk, your club has
promoted safe and healthy commodities. Hereafter you can promote
safe, healthy and favorable electricity by the communal efforts of
paying a small amount—the equivalent of two cups of coffee in the
tea room (see *Hokkaidō Shimbun*, 2 November 1996).

The first citizens' sector wind power station

My suggestion was based on ideas put forward by Shūji Nakagawa,
who initiated campaigns to develop a citizens-owned solar power station in Miyazaki Prefecture in 1994 and in Shiga Prefecture in 1997.
The suggestion was well received by the participants in Hokkaidō,
who felt that they were not making apparent or effective progress
with the usual opposition movement strategies and tactics. A green
power scheme was launched in March 1999 after 30 months of preparation. By this time it was normal for the Seikatsu Club members to
settle their monthly accounts through direct debit transactions from
their banks. In the new green power scheme, the participating members would pay their electricity bills plus a 5% surcharge to the co-op
instead of directly to the power company. The surcharge was to be put
into a trust account held by the Hokkaidō Green Fund, a newly established nonprofit organization. Hence, with one bank authorization,
any member could participate in the scheme 'conveniently, without
difficulty, and with a little power saving'. Non co-op members can
also participate in the scheme by contributing money directly to the
Fund (see Figure 10.1). In a sense, then, the scheme is a means of
fundraising that links donations to an existing direct debit system.

Two different ways of calculating donations were considered: a
fixed payment or a fixed rate. Paying a fixed amount, say, 400 yen per
month or 5,000 yen annually, would be more convenient, but offers
no incentive to reduce power usage to offset the extra cost. Paying a
percentage of one's power bill, in contrast, provides a strong incentive to reduce power usage, as the contribution to the scheme increases or decreases directly according to consumption levels. A fixed rate
system is therefore more readily interlinked with a general power re-

duction campaign. To make the scheme work in this case required close cooperation between the utility, the co-op and the HGF, since the utility had to notify the Fund of participating customers' power charges each month. Fortunately, the director of the sales department of Hokkaidö Electric at the time was agreeable and cooperative in making the necessary arrangements for individual power bills to be sent to the Green Fund.[4]

The scheme initially aimed for 1,000 participants in the first 12 months. In April 2000, 13 months after its launch, there were slightly more than 800, climbing to 1,300 by the end of October 2002. The scheme raised approximately four million yen in fiscal year 2000.

The first citizen-owned power plants in Japan began operating in 1994, beginning with a solar power plant. However, solar has proven to have a higher cost:output ratio, and is therefore less financially viable. In contrast, wind power is an established commercial business around the world (see Figure 10.3).

Since 1997, a growing number of plans for large-scale wind farms have been realized in Hokkaidö. Following a suggestion by a local Buddhist monk and a group of young residents, Tomamae Chö has become a center for wind power generators. At the end of October 2002, 42 turbines with a total output of 52,800kW were in operation including three owned by the town itself and others operated by Tömen Power Japan Co. Ltd. (now, Eurus Energy Japan Co. Ltd.) and Dengen Kaihatsu (the Electric Power Development Corporation, now, J-Power Co. Ltd.). By 1999, wind farms with a total capacity of 500,000kW were planned for Hokkaidö Prefecture.

Hokkaidö has many similarities to Denmark, called 'the birthplace of wind power electricity generation'. Denmark was the first country to ever generate electricity from a windmill, and produces almost 50% of the wind generators in operation worldwide. Both have cold climates, with populations of around five million, similar sized family farms (mainly dairy-farming), and similar GDPs. Akio Ötomo, a research assistant in the Engineering Department at Hokkaidö University and an expert in renewable energy, gave a public lecture in February 1998 entitled 'Make Hokkaidö the Denmark of Japan: Community building with renewable energy', coinciding with growing interest in wind power nationwide. A push to turn the problems of extremely strong winds—including frequent crop damage—into an economic benefit was spearheaded by Tachikawa Machi in Yamagata prefecture, the first place to successfully establish wind farming in Japan as a commercial business in 1996.

Most of the wind farms planned for Hokkaidö were profit-driven, while the others had citizens' sector owners. But the wind in the sky is a 'public' resource that does not exclusively belong to any individual(s) and should not be used solely for profit. The board of the HGF concluded that a nonprofit organization such as its own should therefore become actively involved in power generation as a community enterprise, and decided that this was the best way to invest the funds that had so far been and were continuing to be raised.[5]

After considering various potential locations, the HGF selected a site in Hamatonbetsu Chö on the coast of the Sea of Okhotsk, not far from Horonobe, the site of the aforementioned nuclear facility dispute. Many leading members of the HGF had long histories of commitment to the area. The site was selected primarily for its wind capacity. Another favorable condition is that Tömen Power Japan could be subcontracted to construct, maintain and run the plant. Tömen Power was highly experienced in the wind farming business nationally and internationally. Subcontracting the entire project to them offered reduced financial costs and provided management expertise (thus reducing risk) for the first attempt of a citizen's communal wind power generating project. A 60m 1,000kW turbine was selected at a cost of 200 million yen. A campaign aiming to raise 60 million yen at 500,000 yen/share began in mid December 2000, with the remainder to be borrowed. The campaign generated an overwhelming response—with more than 100 million yen raised in the first six weeks. In the end, 141.5 million yen was raised through the sale of 249 shares to some 200 individuals (124.5 million yen) plus donations from various groups, including the 'Supporter's Club'. 'Hokkaidö citizens' wind electric power generation' was then incorporated, owned by the citizen/shareholders. This was the first company of small share holding by citizens in Japan under a new law. Construction began in March 2001 and power generation began on 15 September 2001.[5] The turbine is expected to generate 2.66 million kW per year operating at 30% of its rated capacity. This is equivalent to the power consumed by 900 households, or 32 million yen worth of electricity. Achieving an output of 30% of rated capacity is high by international standards. Nevertheless, in the second year until the end of March 2003, the new facility came close to this target, achieving 29.2%. The second year dividend of 19,969 yen per unit (500,000 yen), interest of almost 4% a year (0.5% on 5 year fixed term deposits) was paid to the shareholders from the sale of electricity for the 12 months ending March 2003. The importance of the citizen's communal wind power plant, set up by the Hokkaidö Green Fund is that it links the movement and community business through small investments and small share holdings by citizens.

The Hokkaidö Green Fund's achievement was widely reported throughout the country in newspapers, television and other media, receiving general acclaim. It was awarded the second annual 'Environment for Tomorrow Award' sponsored by the *Asahi Shimbun* in May 2001, and in October, the Environment Minister's annual award for the prevention of global warming. It was also cited in the annual 'Environmental White Paper' for 2002.

The Hokkaidö Green Fund's model of using a nonprofit organization to appeal widely to citizens for donations and investment to set up a citizens' communal power plant has since been adopted throughout Japan. Similar projects are under way in five other regions of Hokkaidö, as well as Aomori, Akita, Miyagi, Chiba, Shizuoka, Fukuoka and other prefectures. The Hokkaidö Green Fund is directly involved in some of the projects in Hokkaidö, Aomori and Akita, supporting local NPOs and other citizens. Construction began on the 1,500kW 'Akita Citizens' 1st Turbine' in November 2002 at Tennö Machi, Akita Prefecture. Another new NPO, 'Green Energy Aomori' began construction of a similar 1,500kW wind turbine at Ajigasawa Machi, Aomori Prefecture at the same time. Both began to operate in February 2003 and are producing electricity as planned.

The green power fund: The utilities' version

The activities of the Seikatsu Club Hokkaidö and its spin-off Hokkaidö Green Fund were closely watched by the local branch of the Ministry of International Trade and Industry (MITI: now the Ministry of Economy, Trade and Industry), which made frequent inquiries about membership levels. At the same time, numerous activists were speaking at symposia, public meetings and in the media about the importance of various green power schemes in Europe and the USA. Prominent speakers included Yurika Ayukawa of WWF Japan, Tetsunari Iida, the founder of an NPO named 'the Green Energy Law Network', and myself. Of the major power companies TEPCO was the most eager to create a green scheme, but they were pre-empted by the Ministry of International Trade and Industry, which 'encouraged' all ten of the regional utilities (including Okinawa Electric) to begin similar schemes simultaneously in October 2000.

The utilities' green power fund schemes charged customers a fixed amount of 500 yen/month on top of their power bills.[6] The utilities agreed to match all contributions on a one-for-one basis or more, and commit these funds to developing renewable resources such as wind and solar power. The collaboration between willing customers and the utilities was praised. While each of the utilities' schemes differs slightly, they are all similar in important ways. For example, Tōhoku Electric's fund is managed by a third party foundation, the Tōhoku Industry Revitalization Center (Tōhoku Sangyō Kasseika Sentä). The Center established a special committee, including representatives from two NGOs, to administer the green power fund and make decisions about where the money should be spent. The committee's operations are open to public scrutiny (in contrast to the utilities' and government bureaucracies' 'closed' decision-making processes, see chapter 9). The managing committee publishes regular reports on their website, including total customer contributions, the committee's budget and disbursements, its funding schemes for wind and solar generation and other information about their activities and investments. In Tōhoku Electric's service area, about 8,000 customers signed up in the six months ending March 2001, contributing 9.34 million yen, to which Tōhoku Electric has added 20 million yen from their general revenue.

All nine of Japan's major electric utilities (except Okinawa Electric) and MITI are pro-nuclear and therefore usually directly opposed to the anti-nuclear and environmentalist organizations. These utilities are also often accused of being even more bureaucratic than the many national and prefectural government bodies. Thus the fact that a project initiated by an anti-nuclear nonprofit organization has

Figure 10.1 Flow of green power schemes in Hokkaidō

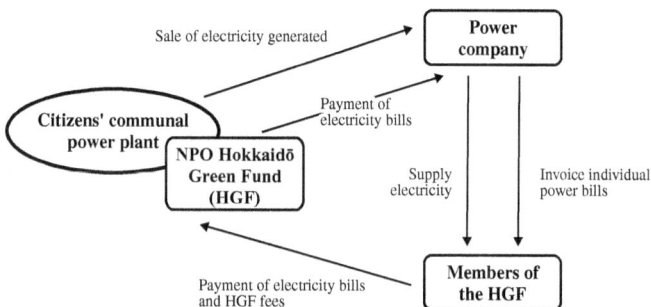

been taken up by MITI and implemented by all ten utilities is truly a monumental event.

Reframing opposition movements

Analyzing the Hokkaidö Green Fund's success

I will now examine the Hokkaidö Green Fund's success using the analytic tools outlined in chapter 5: political opportunities, mobilization structures, and cultural framing. As discussed in chapter 8, the anti-nuclear power movement in Japan can be roughly divided into four different demographic groups: 1) residents near the proposed site, who are primarily the directly affected land owners and holders of fishing rights; 2) residents of neighboring cities and towns supporting the local opposition; 3) non-government/industry professionals in metropolitan areas; and 4) a broader populace in mainly metropolitan areas, primarily housewives alarmed by Chernobyl.

The anti-nuclear campaign by the Seikatsu Club Hokkaidö was initially of the second type, opposing the planned nuclear plant at Tomari Mura and the prospect of the high-level radioactive waste dump site at Horonobe Chö. As mentioned, though, it arose in the wake of Chernobyl, and thus also had the characteristics of the fourth type, enjoying strong support from housewives and the broader community.

When a plan was announced in the summer of 1996 to build a third reactor at the Tomari nuclear power station, the political op-portunities for the opposition movement appeared to be quite limited. Since there were already two reactors in operation, the presence of the power station seemed to be irreversible. It was obvious that op-position to the third reactor would not receive any effective political support. After the 1995 election, the number of seats occupied by the anti-nuclear left parties—Socialists, Democrats including former Democratic Socialists, and Communists—who had supported the referendum bill in 1988 had dropped from 53 to 43 in the 110 seat prefectural assembly. In other words, there was very little prospect of getting a referendum bill passed by the new assembly. Tatsuya Hori, who had previously served as Vice Governor under Takahiro Yokomichi, was elected Governor in 1995. Advocating a 'post nucle-ar future', he had been supported by the Socialists, Shinshintö and Kömei,[7] Rengö (the peak union body) and other organizations. When

the No. 3 reactor was announced for Tomari in 1996, the newly elected Governor vowed to remain 'neutral' and 'make his decision after
listening to the voice of the people of the prefecture'. All the major
parties supported his reelection in 1999, except the Communists. A
lawsuit filed in 1989 seeking an injunction against the Tomari nuclear power plant was finally lost in 1999 when the Sapporo District
Court's rejection of the plaintiffs' case was upheld by an appeal court.
Even without that development, though, the capacity of Governor
Hori, a career bureaucrat in the prefectural level of government, to
resist the political pressure from the national political powerbrokers
and industry bureaucrats was always doubtful. Thus somewhat pre-
dictably, after strong pressure was brought to bear by Muneo Suzuki,
the then LDP national Diet member from Hokkaidō who was later
arrested and resigned following accusations of receiving bribes, and
others, Governor Hori gave his approval for the third reactor in
September 2000. The reactor plan was then proposed to the national
government's Power Development Council for consideration and was
approved.

The anti-nuclear power movement was experiencing tough times
across Japan, even though memories of the accident at Monju in
December 1995 were still vivid. Nevertheless, there were notable
exceptions, such as Maki Machi, Niigata where the opposition to a
proposed reactor won a referendum in August 1996 (see chapter 9),
and Ashihama, Mie Prefecture, where an application for a nuclear
power plant was withdrawn by Chūbu Electric in February 2000,
following a 20 year battle with local fishermen and residents, and the
Governor's final decision to oppose the project.

The anti-nuclear movement in Hokkaidō was also suffering from a
limited capacity to mobilize resources following the long decline of
the labor movement. In the 1988–89 petition campaign, one million
signatures were collected. In a second campaign between September
and December 1998, petitioning for a prefecture wide referendum,
although the target had been reduced to 100,000 signatures, only
97,000 were collected and tabled in the Prefectural Assembly in
December 1998. There was another sharp turnaround in a third
petition for a referendum in April 2000, following an accident at the
JCO plant in Tōkai Mura, Ibaraki Prefecture in September 1999. This
time, the target was once again one million signatures and 780,000
were collected, but as expected the request for a referendum was
rejected by the assembly.

At this juncture a substantial reframing was required for the movement—to present an alternative proposal rather than simply rejecting others' proposed developments. Reducing power consumption, minimizing waste and harnessing strong winds for a community business solution are exemplary types of alternative actions that offer future visions (Hosouchi 1999; Kobe Toshi Mondai Kenkyūjo 2002). The fresh and positive frame signified by the HGF's slogan 'Let's green electricity' appealed to the mass media, and was easier for the general public to accept than negative slogans like 'No nukes'. People who could not be visibly involved in opposition to the new plant, due to employment or other constraints, could support a green power campaign.

An oppositional campaign tends to be a zero-sum game, its success and political efficacy measured in terms of all or nothing—either the project is stopped or it goes ahead. Opponents and proponents are intractably opposed, each casting the other as an enemy with no middle ground for negotiation. The success of a campaign to promote an alternative, in contrast, can be judged in a much more relative manner, as its effectiveness and progress is not determined by extrinsic factors such as the construction of a power plant.

Key members of the Hokkaidō Green Fund have studied the developments in environmental and alternative energy movements worldwide. They have collected, studied and transmitted information on strategies and tactics such as how to implement power saving measures and to nurture solar, wind and other renewable energy resources (Hokkaidō Green Fund 1999). Among other things they have concluded that the terms 'sustainability', 'smaller and dispersed', 'decentralization of power' and 'citizens' initiatives' are not merely fashionable buzzwords, but are also fundamental elements of new electricity policies in European countries.

An important measure of the success of the movement's reframing is that their political opportunities subsequently improved significantly. The Ministry of Economy, Trade and Industry and the utilities had been the enemies of a movement whose identity was defined by an anti-nuclear frame, but they could work with a movement whose objectives were to promote renewable energy, even if they remained skeptical about its effectiveness and economic viability. In a similar act of reframing, the fishermen in Karakuwa Machi, Miyagi Prefecture initiated a campaign entitled 'the woods, the darling of the sea' to reforest the upper reaches of the Ōkawa River. Such

approaches are unlikely to create enemies for the green power movement. Opponents find it difficult to counteract strategies that have
demonstrably broad public support both within Japan and globally.

The oppositional type anti-nuclear movement has had almost no
success in expanding its support base in recent years, largely due to
its very limited political opportunities. In contrast, the HGF has found
many supporters and collaborators outside of its traditional support
base. For example, we have already mentioned the cooperation with
the Hokkaidö Electric Power Company in the HGF's billing system.
Another example, is Tömen Power Japan the company subcontracted
to construct, maintain and operate the turbine. The media was also
helpful in reporting the Green Fund's progress and defining its
importance. The HGF's success also helped to consolidate or reinforce
the network of activists, scholars, environmental NGOs and residents'
groups throughout Japan.

Analyzing the utilities' green power fund schemes

Here I will analyze the utilities' green power fund schemes in terms of
the three dimensions discussed in chapter 6: (1) the decision-making
process, (2) technical methods, and (3) values.

As mentioned, TEPCO's interest in what was essentially a citizens'
initiated project was monumental. That the program was embraced
by MITI and, at MITI's instigation, introduced by all ten of the major
utilities is truly astounding, especially considering that the electricity
industry is one of Japan's most closed and rigid public policy areas. We
should note, however, that there has been some suggestion that MITI's
true motives for embracing and promoting the scheme remain hidden.[8]

The 'top-down' introduction of green power fund schemes implies
several important changes at the decision-making level. First, the
Ministry and utilities validated the framing of renewable energy as
'green power'. Second, the fact that more than 20,000 households
nationwide joined the ten utilities' green power fund scheme in the
first six months directly challenged the power industry technocrats'
long held views of consumers as 'passive, easily manipulated by
media, self-interested and largely ignorant of and uninterested in
energy issues'. The industry's adoption of similar schemes indicates
that the technocrats' perception of consumers through the lenses of
mass society theories, 'Olson's free riders' (see chapter 4), or C.W.
Mills' 'cheerful robots' can be challenged and altered. Third, the
funds' administration through third-party committees that includes

citizens' representatives and make all information publicly available entailed what were once incredible concessions from this very closed and rigid bureaucracy to the growing demands for a self-determining civil sector. Even the schemes themselves, enabling ordinary citizens to opt-in to a scheme with a mere 500 yen/month contribution—in effect a vote for green power—entailed a previously unthinkable opening of the industry's policy-making processes.

At the same time, the industry's primary focus remains fixed on large-scale fossil fuel and nuclear powered generators—and thus this new found interest in renewable power sources is not entirely convincing. In other words, their efforts to promote green power are lost in mixed messages, in contrast to the HGF's clear and unambiguous slogan 'Let's green electricity'. This ambiguity is reflected in the limited number of people who have joined the utilities' green power fund schemes. TEPCO has the largest number of participants, while Tōhoku has signed-up the highest proportion of their customer base. Tōhoku's success in this regard might be attributable to the fact that most of the money collected through their green scheme is allocated to purchasing electricity from the wind turbines operating in their local region. The other utilities have received lukewarm responses to their less enthusiastically promoted schemes.

Another significant problem is the way the collected funds are to be spent. In a bidding system that favors the buyers, the utilities, the intention is to add 0.5 yen/kW to the utilities' bidding price when buying power from wind farms with a generating capacity in excess of 2,000kW. The money collected by those utilities with few suitable locations for wind farms is to be directed to Tōhoku Electric, which has many suitable locations. Whether the system contributes in any way to the development of additional green power facilities is doubtful. The 'greening' effects of these schemes on Japan's power industry have been relatively less substantial than what has been achieved by the compulsory premium price wholesale electricity systems introduced in Germany, Denmark and elsewhere.

From the start, although the scheme's transparency and objectives have been deservingly applauded, its effectiveness in promoting the widespread use of renewable energy was questioned. All of the utilities' schemes had more or less stalled by 2002, with hardly any growth in participation (see note 7). It seems that the donation collection system has reached its limit.

Various green schemes

From the consumers' perspective, there are at present five different types of green power schemes: 'green fund' (opt-in surcharge), 'green investment', 'green pricing' (commodity), 'green certificates' and 'green taxation' (ear-marked taxes).

In a 'green fund', an opt-in surcharge system, as discussed above, consumers can choose to pay a premium for their electricity—either a fixed payment or a percentage of their electric bill—to finance green power generators and operators. These systems are easy for consumers to comprehend and commit to. In Europe and North America, an average of 0.5–1% of the grid connected customers take part in such schemes. In Freiburg—the well known eco-conscious city in Germany—the participation rate of around 10% is especially high. A 'green investment' scheme such as the Hokkaidö Green Fund differs markedly from the surcharge system in that the investors become shareholders who may receive a financial return (dividends) on their investments. A 'green pricing' commodity scheme can only work in a liberalized/deregulated market, where utilities can market more expensive green power as a distinct commodity which customers may purchase if they so choose. In California, where the market was opened in April 1998, there are various green power commodities being marketed on the basis of having been 50% to 100% generated from renewable energy resources.⁹

The scheme that has attracted the most international attention recently is the fourth, the 'green certificate' system. Simply put, when imposed by national governments across the board, a minimum quota is imposed on the utilities; say for instance, 5% of all the electricity they distribute must come from renewable sources. Tradable certificates can then be introduced to complement any shortfall, since suitable locations for wind turbines, the primary commercially viable renewable power source, are limited. If a utility can only generate or purchase 3% of its total output, it must then purchase green energy certificates in the market to meet its shortfall. This system was first introduced on a large-scale by the Dutch government in 2001 (see http://www.certiq.nl/; http://www.recs.org/), followed by the Swedish government in 2003. Although still in an experimental stage in Europe, the Japanese Ministry of Economy, Trade and Industry is seriously considering something similar for the corporate business sector.¹⁰

The fifth type of green power scheme, 'green taxation', entails earmarked taxes, in which funding for developing renewable energy resources is raised through a government tax or levy on power bills. Environmental taxes and carbon taxes, if earmarked for this purpose, can also be included in this category. The German city of Aachen attracted international attention in 1995 when it introduced a system that included a 1% increase in electricity prices, with the money earmarked to pay for the higher cost of renewable energy. The city is committed to paying higher prices to solar and wind power producers at 'full cost rates' (FCR) of nearly ten times higher for solar power and 1.3 times higher for wind power than for power from non-renewable sources for the next 20 years. This FCR system has come to be known as the 'Aachen model', which was effectively implemented across Germany when it was incorporated into the 'Renewable Energy Law' in April 2000.[11]

The most important factor about any of these schemes is that they provide conclusive evidence that clearly expressed consumer demand is heard by governments and utilities, who in turn have responded with various efforts to develop renewable energy. In that sense, green power schemes are open systems developing through trial and error, where initiatives and ideas from NGOs can become official policy, in contrast to the closed and conflict-ridden relationships between opposition movements and governments (Green Energy Law Network ed. 2000).

The mutual transformation and permeation of government policies, social movements and businesses

The green power scheme and changes in nuclear energy policies

The green power scheme is a prime example of an increasing number of environmental policies and programs that were first initiated by environmental NGOs, implemented on a small scale by local authorities, then taken up nationally before spilling across the borders and being adopted by many other countries.

Figure 10.2 charts the chronology of the development of green power schemes in California, Germany and Japan. These developments are closely related to changes in nuclear power policies. As previously discussed, the world's first green power scheme, 'PV Pioneers' was introduced in 1993 in Sacramento, California by a

Figure 10.2: The development of green power schemes in California, Germany and Japan

California	Germany	Japan
Nuclear power plant shut down by referendum (June 1989) ↓ Promote solar energy ↓ 'PV Pioneers' program started by SMUD (1993) ⎮ ← California opens its ⎮ electricity market ↓ A variety of 'green electricity' schemes introduced in California ⎮ ← USA opens the ⎮ electricity market ↓ Green electricity schemes introduced nationwide	Forests damaged by acid rain (early 1980s) ↓ Growing awareness of environmental issues ↓ Chernobyl Incident (April 1986) ↓ SPD decides to abandon nuclear power plant (August 1986) ↓ End of Cold War Nuclear fuel reprocessing plant & high breeder reactor cancelled and all reactors in former East Germany closed down for failure to meet the safety standards of West Germany (1989–91) ↓ Growing awareness of global warming issues ⎮ ← Popularizing wind power ⎮ generators in Denmark ↓ Begin purchasing renewable energy by fixed price system (1991) ⎮ ← Municipal authorities ⎮ introduced measures to ⎮ promote renewable ⎮ energy in Aahen and ⎮ Freiburg ↓ COP1 in Berlin, surge of wind power generators, RWE launches 'green electricity program' ↓ The 1998 Red-Green coalition agreement to close all nuclear power plants nationwide ↓ 'Renewable Energy Law' Enacted in April 2000 ↓ Agreement reached between government and utilities to phase out all nuclear plants (2000)	*A Choice for Post-Nuclear Society* published (July 1996) ↓ Majority of residents reject proposed nuclear power plant in the nation's first referendum in Maki Machi (August 1996) ↓ I explained SMUD's 'green power' scheme in Sapporo (October 1996) ⎮ ← Tomari N-Plant issue ⎮ ← Horonobe issue ⎮ ← Wind power boom in ⎮ Hokkaidō ⎮ ← The NPO Law enacted ⎮ (December 1998) ⎮ ← Citizen's communal ⎮ solar power projects 'Green power movement' launched in Hokkaidō by the Seikatsu Club co-op (March 1999)—800 households joined ↓ Hokkaidō Green Fund established (July 1999) ↓ The first meeting of Diet members to promote renewable energy (Nov. 1999) ↓ Intention to table 'Green Energy Law' abandoned (April 2000) ⎮ ← MITI retains its policy ⎮ initiative Power companies introduce own versions of 'green power fund' (October 2000) ↓ The nation's first citizens' communal wind power generator (by HGF) commences operation (Sept. 2001)

public utility that had been forced to shut down its nuclear power plants by a citizens' initiated referendum. Other green power schemes sprang-up across the state, then across the nation as the electricity industry was deregulated.

The residents of Germany have very high levels of environmental awareness, prompted by severe acid rain in the 1980s, the fallout from Chernobyl and a variety of pollutants in the former German Democratic Republic and Eastern Europe. It has adopted many advanced policies from its neighbors—especially the Netherlands and Denmark—and leads the EU's environmental policies. Since the end of the Cold War in 1989, there have been major changes in its nuclear power policies. The agreement between the German government and the utilities in June 2000 to scrap all 20 of their nuclear power plants after 32 years of operation was a landmark in world energy policies. The agreement was formally signed in June 2001, and the Atomic Energy Act was amended accordingly in February 2002, taking effect in April. The amendment to the Act clearly shifts the policy focus and purpose of the Act from the promotion of nuclear power to 'phase out the use of nuclear energy for the commercial generation of electricity in a structured manner, and to ensure on-going operation up until the date of discontinuation'.[12]

Because of Germany's decentralized governmental system and the strong presence of Green Party members and other environmental political forces in local governments, numerous municipalities and regional governments have been able to initiate various attempts to develop and promote renewable energy resources, with strong support from local NGOs. As indicated in Figure 10.2, since the Act on Feeding Electricity Generated from Renewable Energy Sources into the Grid (Feed-in Law), which became effective in January 1991, stipulating that the utilities must purchase renewable energy at higher prices, the number of wind turbines in operation has surged. At the end of 2003, Germany was home to more than one-third of the world's commercial wind turbines, with a total rated capacity almost 30 times greater than Japan's (see Figure 10.3).

The relationships between the introduction of green power schemes and changes in nuclear power policies vary widely from country to country, and are therefore not easily generalizable. Nevertheless, some general trends in these relationships are discernible. Figure 10.4 illustrates the process of change through five stages.

The process began with growing public awareness and opposition movement activities following the Three Mile Island (1979) and

Figure 10.3: A comparison of the accumulated capacities (kW) of the world's major wind-powered nations and Japan

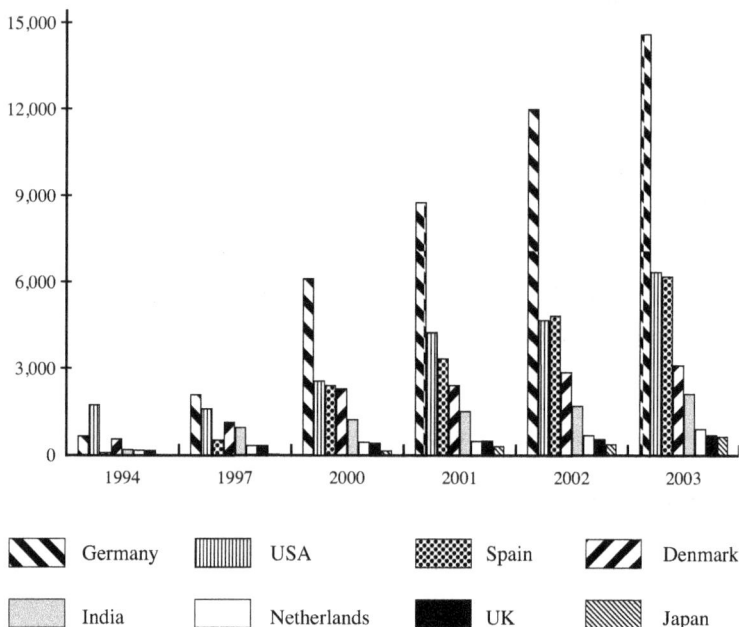

Note: Figures are for the end of each year. All data is from the Windpower Monthly except for the 1994 data, which is from the US Wind Power association

Chernobyl (1986) reactor incidents. Governments began reviewing and eventually scrapping programs to extract plutonium from spent fuel rods (in the USA, for example, the review began in 1977 and the program was scrapped in 1983, the German program was scrapped in 1989, etc.).

The termination of nuclear fuel reprocessing programs was the beginning of across-the-board changes in direction across all nuclear power programs. The second stage occurred at the same time, as environmental NGOs and local authorities began various experiments to promote the use of renewable energy, including the earliest green power schemes. These local initiatives were then taken up by central governments and implemented nationwide. Encouraged by this official endorsement of their efforts, the local authorities began to introduce even more progressive measures. There were numerous example of this process in the USA and Germany in the mid 1990s.

The third stage began when power generated by nuclear power plants became more expensive than coal or natural gas fired thermal power generators, and plans for new reactors stalled, for both economic reasons and because of the social risks. With the cancellation of new reactor proposals, the nuclear power program effectively came to a halt.

The fourth stage involves phasing out aging and trouble ridden nuclear reactors, and the deregulation of the electricity industry (USA in 1989 and Sweden in 1999). The fifth stage involves making arrangements for phasing out the entire nuclear power sector. In order to legally require the termination of privately owned nuclear power plants, the governments and owners must reach agreement about financial compensation and timing etc. As mentioned, the world's first such agreement was reached in Germany in June 2000. The biggest hurdle in this instance was how to ensure an adequate return on the utilities' investments. In the end, both parties agreed that each plant would have an operating life of 32 years from the time it commenced operations, and the industry would not receive any financial compensation from the government.[13]

Following the accidents at Monju in 1995 and at the JCO plant in 1999, Japan might be considered to be in the second stage of this process, but its government is still firmly committed to continuing its reprocessing program.

Figure 10.4   The relationship between green power schemes and changing nuclear power policies

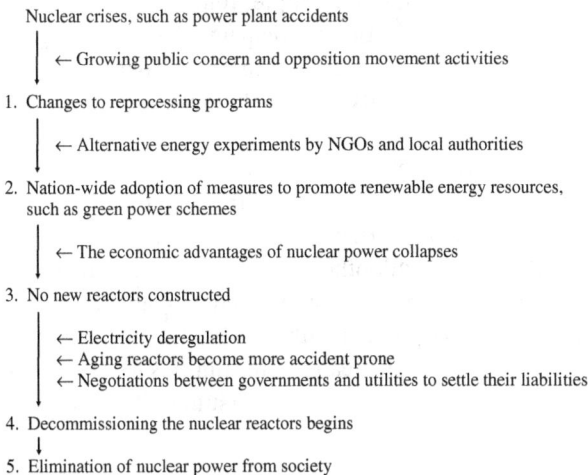

Nuclear crises, such as power plant accidents

⎮   ← Growing public concern and opposition movement activities
↓

1. Changes to reprocessing programs

⎮   ← Alternative energy experiments by NGOs and local authorities
↓

2. Nation-wide adoption of measures to promote renewable energy resources, such as green power schemes

⎮   ← The economic advantages of nuclear power collapses
↓

3. No new reactors constructed

⎮   ← Electricity deregulation
⎮   ← Aging reactors become more accident prone
⎮   ← Negotiations between governments and utilities to settle their liabilities
↓

4. Decommissioning the nuclear reactors begins
↓

5. Elimination of nuclear power from society

Collaborative environmentalism

Led by the various international and domestic NGOs, environmental movements have increasingly changed tack from zero-sum
'conservation' strategies of criticizing and demonizing the industry
and authorities, to more collaborative strategies that lead to greater
direct participation in shaping policy, working with authorities and
businesses while still maintaining a critical position. In this sense,
the citizens' sector in Japan and elsewhere is undergoing a substantial transformation.

'Collaboration' is a key element in the fluid network oriented
contemporary societies. Collaboration is a commonly used word
in English, meaning 'working together'. According to the *Oxford
English Dictionary*, it tends to mean to 'work in combination
for a particular purpose with those with little direct connection'.
Since 1996 I have argued that the word should be defined as 'a
limited purpose ad hoc coalition or partnership of equal partners'
(Hasegawa 1996a: 249). Simply 'working together' or 'a working
relationship' are not sufficient. A collaboration must have 1) an
equal partnership or horizontal relationship stressing 'equality', 2) an
aspiration to achieve shared targets, 3) interorganizational/sectoral
and interdisciplinary participants and 4) a specific purpose. A
collaboration is an interorganizational/sectoral and interdisciplinary
coalition in which people from different occupations, different
places, and different groups work together in a non-routine manner. It
overrides longstanding social norms of clear institutional boundaries
between businesses, NGOs and government bureaucracies. Whereas
a partnership, like a married couple, tends to mean an enduring
relationship, in a collaboration participants freely join together in a
non-binding relationship to achieve common goals, each assessing
the achievements and merits of the relationship independently and
on their own terms.

A collaborative strategy, I argue, is especially useful and
suggestive for social movements in the Japanese context today,
where the government policy structures remain closed and rigid.
Meeting and working with others can stimulate creative and lateral
solutions to the problems being addressed. Collaboration is a model
for a decision-making process with greater citizens' participation in
a more open system, which creates opportunities for the development
of trust between the different parties pursuing different agendas.

Collaboration only works where there is mutual desire and commitment to finding common ground. A society's capacity to expand its network orientation is also a prerequisite, where mutual trust can develop through networking, exchange and socializing. The idea of a collaborative society presupposes a radical transformation of the 'public sphere', from being almost indistinguishable from 'official' and 'governmental', to something much closer to 'public' in the true sense. As I argue in chapter 11, a new public sphere is emerging in Japan.

New social experiments from collaboration

Prompted by the threat of global warming, another 'energy revolution' is presently underway worldwide. Reducing demand has been given the highest priority. Where energy must be used, its most efficient use is considered. Instead of oil, coal or nuclear fuels, the more benign natural gas is increasingly being used. Smaller and more decentralized generating facilities are becoming more common, employing cogeneration and renewable resources such as solar, wind and biomass. More environmentally friendly automobiles are being developed, using natural gas or electricity or a combination of fuels (so-called 'hybrid' cars). In many regions there have been marked changes in transport policy from automobiles to public transport (for example, new light rail tram services). The force driving this energy revolution comes from collaborations between environmental NGOs, electric utilities, the business sector and government bodies (Iida 2000; Imaizumi 2001).[14]

In the US, numerous solar and wind power projects are developing through various collaborations between the US Department of Energy, state governments, utilities and NGOs. In Germany, Denmark and the Netherlands—the world's leaders in policies to reduce energy demand and to promote the use of alternative energy sources—progress has been and continues to be made through multilayered collaborations involving regional governments, utilities, private research institutes and environmental organizations. In Japan too, more and more wind farms are being constructed through collaborations between regional governments, local groups, scholars, businesses and financial institutions, as in Tachikawa Machi, Yamagata Prefecture, Tomamae Machi and Kitahiyama Machi, Hokkaidō Prefecture. A large-scale wind farm is being built on the Shimokita peninsula in Aomori

Prefecture, in close proximity to a nuclear fuel processing facility (see cover photo and Figure 9.2).

It is counter-productive in contemporary society to treat business development and environmental conservation as mutually exclusive or diametrically opposed. At the celebration of the Earth Day's 25th anniversary in 1995, one of the early leaders was quoted as saying that there were now an amazing number of environmental businesses created by collaborations between environmentalists and business people—some between former enemies. This, he said, was unthinkable in the early 1970s (*Los Angeles Times*, 30 April 1995).

The environmental business councils of Europe provide prime examples of such collaborative interactions with the business sector. For example, the European Business Council for a Future with Sustainable Energy (Council e[5]) was founded in Brussels in 1996. The name e[5] expresses the aims of the council: to promote a sustainable energy future in five areas: e-nergy, e-nvironment, e-conomy, e-mployment, e-fficiency. At the Kyoto Conference to Prevent Global Warming these 'environmental business NGOs' were as prominent as the international environmental NGOs such as Greenpeace, Friends of the Earth and the WWF. It seems clear that the best chance of preventing global warming lies in more efficient uses of energy, which generates countless business opportunities. Perhaps in the 21st century environmental businesses will join welfare and information to become the core industries that steel, shipbuilding and car making were in the 20th century.

Businesses, environmental NGOs, consumers and governments have common interests in nurturing environmental businesses. These shared values and objectives provide the basis for future

Figure 10.5: Interpenetration of government, business and citizens' sector

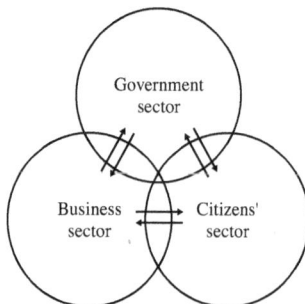

collaborations among these groups. As illustrated in Figure 10.5, the institutional barriers between the government, commercial and citizens' sectors have been more rapidly overcome in the EU countries, enabling greater interpenetration of these sectors. For the government sector, collaboration with the environmental movement means involving them in the policy-making processes. It has become increasingly apparent that the most effective way to educate the public about the government's existing and proposed environmental policies is to enlist the environmental movements to the cause. This, however, requires that the policies being promoted are acceptable to the environmentalists, which in turn requires that the policy-making processes are open to the political ideals and ideas of the environmentalists. Only when the two parties have reached common ground can a joint public education campaign be organized.

For the environmentalists, collaboration means that they can no longer adopt an 'outsiders' position of simple criticism and opposition to government and business development proposals. Instead they must pursue alternative development objectives of their own, collaborating with government and business where possible while maintaining their independence. This 'insiders' approach appears to be the only practicable means of realizing alternative ideals through policy and concrete measures. The activists' alternative ideals will only be deeply inscribed in their society when they have been endorsed and validated by incorporation in government and industry policy. Demonstrating the public demand for these alternatives by successfully marketing them through business enterprises is one of the most effective means of convincing the policy-makers to change in this direction.

At the same time, environmental movements that have become large international NGOs must invest enormous amounts of energy into their own financial wellbeing. Organizations such as the WWF and Greenpeace therefore operate a number of businesses towards financing their core activities. Greenpeace Germany, for example, owns and operates an independent electricity wholesaler of green power products for private and corporate customers (as described in chapter 12). The success of the Hokkaidō Green Fund was a major turning point in Japan—the movements' initiatives became official policy and various new enterprises were born.

For the business sector, as consumers become increasingly eco-conscious, an eco-friendly public image is almost a necessity and, as mentioned, new business opportunities and markets are opening in environmental fields. Large corporations such as Sony and Asahi Brewery do not purchase 'green power certificates' only to be doing their duty as global corporate citizens but to be seen to be environmentally conscious and thus enhance their public image. In some respects then, business and movement objectives are at least in part beginning to overlap, and some parts of the business sector are becoming significant parts of the environmental movement. Of course there are still very powerful conservative strongholds in the business sector, such as the Federation of Economic Organizations (Keidanren), one of Japan's peak business bodies, and the oil industry who generally oppose the environmental movement.

The government and business sectors in developed countries typically have had very long and intricate relations. Political-economy debates have long been concerned with the merits and characteristics of private vs. public enterprises, market vs. planned economies. Now this one-on-one relationship includes a third party, the citizens' sector, as illustrated in Figure 10.5. The inclusion of the citizens' sector as a partner in collaboration has been the driving force behind the further development of a new public sphere and the second wave of environmentalism worldwide since the 1980s (see chapter 1 for discussion of the first wave environmentalism of the 1970s).

In the 21st century, environmental movements, NGOs and NPOs will be judged by their abilities to influence and affect policy. The relationship between the environmental movement and official policy is clearly becoming ever more dynamic.

11　Civil society and the new public sphere[1]

Reinstating the public: Preface

From its beginnings, Western sociology has focused on communality (communal relationship) and publicness (public space; public sphere) as the fundamental principles of order in a modern civil society. Although in Japan, sociology has for a long time paid very little attention to matters concerning the public, there was a sudden and rapidly increasing number of articles on the public and publicness in social science research fields, including sociology, from the mid 1990s, as shown in Figure 11.1. At the same time there was a parallel surge in the number of articles on civil society. These are typical of the drastic changes in the Japanese conceptions of publicness—a move from a state-centered focus to a focus on civil society (see Yamaguchi et al. eds, 2003).

A search through the 'Bibliography of Japanese Sociology Database' reveals that communality has often been discussed in studies of local community, both urban and rural, generally at an empirical level of analysis (Yoshihara 2000). In contrast, most of the papers on publicness or the public sphere are more recent, and following Habermas, are almost wholly theoretical. Notably, as of 20 January 2000, the Bibliography did not have a single reference to the phrase 'public philosophy' (kōkyo tetsugaku).

Sociology is typically defined as 'a social science research field studying social phenomena from a perspective focusing on communal human life' (Aoi 1993: 599). Communality—the sphere of communal life, social relationships, or social solidarity—may well be seen as the defining object of sociology as a social science, in much the same way that the study of market mechanisms define economics and the analysis of policy-making and the exercise of public power defines political science. Consider, for example, Durkheim's Suicide: A Study in Sociology or Parsons' analysis of the Hobbesian Problem of Order.

Communality, possibly because it seems such a common word in the Japanese context, did not merit an independent heading in any Japanese dictionary of sociology until quite recently, when my proposal as one of the editors and my article as a contributor to

Iwanami Lexique of Sociology were accepted, and communality was finally included as an independent subject (Miyajima ed. 2003). The term 'public' is highly polysemeous, highly contested, and as we shall see, is often manipulated for political purposes. Here, 'public' will be defined as having the nature and values belonging to the public space. The term 'public' clearly has an oppositional relationship to the 'private', and because the borders between these two realms are relative, so too are the meanings of the terms. The dominant interpretation of the 'public' in any society largely depends on cultural and political traditions.

This chapter will examine five interrelated aspects of publicness. The first is to recognize the polysemia of the concept, which differs widely historically, cross-culturally, and intra-culturally. Focusing on the more significant of these interpretations, the second aspect is the concept of the public as the unifying principle for civil society. Bellah and others have pointed to a looming crisis in advanced societies, in which excessive individualism may be destroying the very foundations of individualism itself. In this context a revival of public philosophy is being sought.[2] Third is the concept of the 'public sphere' following Habermas. Fourth is a concept of the 'public good' as a standard for evaluating public policies. Fifth is the most recent

Figure 11.1   Number of articles including 'publicness' or 'civil society' as key words

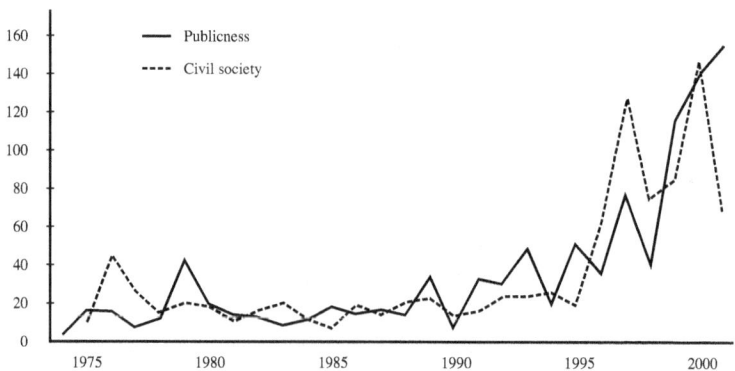

Data source: Index of Journal Articles in Japan (until October 2002). Adapted from Yamaguchi et al. eds (2003: 2) with permission.

formulation of the public, manifest as the non-profit, non-government 'citizens' sector'.

In the English-speaking world, following the end of the Cold War and the 'democratic revolution' of Eastern Europe, there has been a gradual resurgence in theories of civil society beginning with Arato, Cohen and others (Cohen and Arato 1992). To begin with the conclusion, the issue of communality (social solidarity) and the public in today's terms is expressed in the following questions: 1. how to form, maintain and reproduce a unified and civil society, 2. how to ensure transparency and explanatory responsibility in public matters, social justice and efficiency, and to maintain appropriate standards of public services, and 3. what sort of being or entity is this 'citizen' who is communal, and comprises the public of the new public sphere? Of course, these are very practical, policy-theoretical questions, and not limited to sociological discussions.

Changing concepts of public

What is public?

It is useful for clarity to begin by noting that there are significant differences between the typical Japanese and English understandings of 'the public'. In the West, simply put, 'public' means 'widely open to the society' (Hashizume 2000: 451) or 'open to anyone'. While it may sometimes include the government (as in 'public works' or 'public services') it can also signify an other who is opposed to the government. The public is almost never identical to the government in English, but this is the association that Japanese people most often make. In English the difference between the words 'public' and 'private' is quite unlike the difference in Japanese, where they correspond to dichotomies such as above and below (in social status), superior and inferior, and positive and negative. In English 'public' signifies many unspecified people, while 'private' refers to close people like family and friends and specially designated relationships, such as those with employees and shareholders. That is, public is all inclusive while private is exclusive. Of course, both in practice and theoretically, the boundary between the public and private is both plastic and permeable.

It is worth stressing that this English concept of public has had very little usage in Japan, where 'public' and 'private' have long referred to the government/administration and private enterprises respectively (Hasegawa 1998d). The Japanese reading of the Chinese character for 'public' is *ooyake*, which also points to the emperor, emperor's court, government or politics. The Japanese translation of the English term 'public interest' *köeki*, literally means 'benefit to many unspecified people', but a very Japanese bias can be seen in dictionary definitions such as the 4th edition of *Köjien* (1991), Japan's most popular dictionary, which defines *köeki* as 'benefit to either the State or society and public'. This clearly expresses the long-standing belief in Japan that the public interest must be defined and realized by the government, in particular the central government.

Further evidence of the peculiarities of the Japanese use of this concept can be seen in the debates leading up to the enactment of the Law Concerning the Promotion of Specific Non-Profit Organizations Activities (the NPO Law), promulgated in March 1998. One of the biggest points of dispute was whether to use the word *köeki* ('public interest') in the necessary requirements for defining organizations subject to this law, with the government in favor and the citizens' movements that had initiated the Bill staunchly opposed. Finally the word *köeki* was dropped and the full expression 'interest of many unspecified people' was adopted. That the citizens' groups, the strongest proponent of this law, refused to be beholden to the 'public interest' implied by the term *köeki* demonstrates how eristic and confused this concept can be in Japan (Matsubara 1999). While the citizens' groups won that point, they lost others. For example, they had lobbied to name the law the 'Citizens' Activities Promotion Law', but that was not acceptable to the ruling Liberal Democratic Party. There is a provision in Article 1 that refers to 'specific non-profit activities as a form of free social contribution activities carried out by citizens, including volunteer activities'. This rather clumsy phrase contains a significant landmark in the Japanese move towards a civil society, as the only article in Japanese law to use the word 'citizen'. Its inclusion was also hotly debated in the process of enacting this legislation, and further illustrates the unique Japanese connotations of these words.

As mentioned, Japanese sociology did not actively discuss 'the public' or 'public interest' until the 1990s, in part perhaps because of the imperial connotations of the word ooyake. Hashizume (2000) argues that this concept of the public is not, however, uniquely Japanese, but rather a Hegelian dialectical opposition between the State and civil society, and is common to 'late-modernizing' countries such as Japan and Germany where modernization was a top-down project of a strong centralized government. But in contrast to Japan, after World War II, a strong decentralized system was intentionally adopted in Germany in order to weaken the national government. As discussed in chapters 10 and 12, this system eventually proved to be highly conducive to the initiation of a variety of progressive environmental policies at the local level.

Interpenetration of public and private

For the remainder of this chapter I will employ the term 'public' in some of its many English senses, and discuss its transformation in universal terms. From this perspective it is apparent that in recent decades the public has substantially changed in many dimensions.

One dimension is addressed by Habermas's thesis of the 'colonization of the life-world' (Habermas 1981), where the life-world— the private sphere—is being encroached upon and hollowed out by the public sphere, in the forms of the welfare state, an all-pervasive market economy, techno-scientific rationalism and social engineering, and the machinations of the mass media. This process took on a new form in the 1980s and 1990s as the privatization of public enterprises further encroached on the border between the 'public' and 'private'. Spearheaded by the political leadership in the UK and USA, this trend rekindled debates around the world over whether public properties should be maintained and public services provided 'publicly', funded by government taxation, or 'privately', as commodities provided by private enterprises seeking profit. This fundamental debate, found in all introductory economics textbooks, is becoming increasingly confused, as the balance has increasingly shifted towards the private. The third pathway, discussed below, of supplying public goods via the citizens' sector NPOs is rapidly increasing both domestically and internationally.

At another level, the public space has been radically transformed and expanded by the forces of globalization. The geo-political borders that define nation-states also separated and differentiated national publics, while excluding non-national others from this public sphere. The modern nation is, in principle, a space equally open to all of its nationals, and the rights of these nationals are the rights that the State confers and guarantees. The raft of phenomena referred to as globalization have effectively been an unprecedented and continuous assault on the institutional and geographical barriers between nations. The effects of globalization are various and heterogeneous, but the term invariably refers to the rapid disappearance of obstacles to spatial movement, both for physical movement and more especially information communications. As these barriers have disappeared the idea that the earth's population is a single public—a view long espoused by the environmental and peace movements—has become increasingly shared around the world.

As the institutional and ideational boundaries between nation-states have become more vague and permeable, so too the distinctions between their publics—and vice versa of course. At the beginning of the 21st century, people find themselves living in a complex, multi-layered public space in which they have different degrees of power and responsibility at each level of community, district, nation, region, or earth. The increasingly shared view that each of us is connected to all of these levels of the public is accompanied by an increasingly shared value—the belief that this new public sphere must be open to everybody.

Should trees have rights?

Claiming that public rights are open to everyone, however, does not resolve the questions of inclusion. In recent years the supposedly self-evident postulate that only human beings have rights in the public realm has been repeatedly challenged by philosophers, animal rights activists, environmentalists and in the courts. Stone presents an excellent argument for claiming that trees have the legal rights to stand as plaintiffs in 'Should Trees Have Standing?' (Stone 1972).[3] In Japan, lawsuits have been filed claiming legal rights for various species of non-human nature, as in the Amami rabbit Pentalagus furnessi lawsuit (filed 1995), with four species as plaintiffs as well as some people (Shizenken Seminä Hökokusho Sakusei Iinkai (Environmental Rights Seminar Report Committee) 1998). The idea of civil or public rights can theoretically be extended to rare organisms or organisms in general. Furthermore, the World Heritage List's campaign for the preservation of historic streetscapes and local heritage trusts have proposed that certain rights be accorded to various non-creatures, scenery and cultural objects—specifically a right to be preserved. Theorists of the public and the public sphere are being forced to reconsider and revise their anthropocentric perspectives. Here it is worth recalling Dunlap's original vision for environmental sociology, which was to bring an end to sociology's inherent anthropocentrism (Catton and Dunlap 1978 and chapter 1 of this book).

Reviving public philosophy: The free rider problem

When considering communality and publicness, Olson's 'free rider problem' must not be ignored (Olson 1965 and chapter 4 of this book). Ideally, a civil society is one based on voluntary contracts made by free individuals. In other words, self-determination and voluntary respect of contracts are the basic values of a civil society. This raises a question that has plagued modern philosophy from the beginning—with just freedom and volunteerism as postulates, how is it possible to maintain social order? Hobbes' answer was the investment of individual sovereignty in the Leviathan (State), which entails surrendering to limited freedom by a voluntary social contract while reserving the right to resist.

Consider this from the Olsonian perspective. Doing volunteer work for an environmental cause and going to vote are examples of 'cooperative behaviors' in the public realm. Not participating or not going to vote are then 'non-cooperative behaviors'. Co-operative behaviors may involve costs for the individual, while non-cooperative behaviors may be cost-free. To the question of why it is difficult to increase cooperative behaviors, Olson answers that people are self-interested egoists, who avoid the costs that accompany cooperative behaviors whenever possible while continuing to enjoy the benefits. In short, they are free riders.

Most of the advice available about how to encourage cooperative behaviors emphasizes the importance of enlightening the public through activities such as providing information, public relations, and education, but these approaches all assume some sort of pre-established harmony thesis. According to Olson, even when people recognize a 'common interest' or 'just cause'—that is, communality—this does not immediately motivate people toward cooperative behaviors. A comfortable environment, a stable political system and social order are in principle open to every member of a society. These are 'non-exclusive' collective/public goods. If, as Olson argues, people behave to maximize their individual profits, it follows that they will attempt to enjoy their share of public goods such as a 'comfortable environment' and 'stable political system' without personally bearing any of the costs in time and labor. The problem raised by Olson is that without any special conditions, nobody except the good-natured will contribute.

Olson argues that cooperative behaviors among selfish people will only be realized under special conditions. These conditions are summarized in chapter 4, so here we can skip to the conclusion. Since these conditions are coercive or small in scale and both violate the basic postulates of a civil society, Olson advocates offering 'selective incentives'.

A selective incentive is a reward proportional to the degree of contribution, such as wages or salaries, but unlike business or government sector workplaces, in the public space of civil society it is difficult to offer financial incentives because the 'goods' being produced do not necessarily have market value. Resource mobilization theories therefore argue that the only way to expand voluntary cooperative behavior is to offer 'purposive incentives'—a sense of personal satisfaction or achievement in having contributed to a meaningful objective—and 'solidarity incentives'—such as the self-esteem and satisfaction derived from working in cooperative relationships with others.

Paths to the 'public': Self-determination

An important dimension to issues of communality and publicness is 'self-determination'. Demands for self-determination have fueled the recent expansion of movements demanding public referendums in Japan, and more generally are a defining characteristic of the new social movements (see chapters 5 and 8). 'Self-determination' has been a defining objective of feminist movements and movements seeking equal recognition for minority and indigenous cultures. It is also seen in movements demanding various rights to 'freedom of expression' such as the decriminalization of 'recreational' drugs, prostitution and gambling. Many arguments for the legalization of euthanasia also rely upon the individual's right to self-determination.

In other words, issues of self-determination are central to a vast array of the forces for social change that are reshaping contemporary societies, and are in fact defining characteristics of modernity itself. Which brings us again to the problem of social order: under what circumstances and to what extent can the imposition of restrictions on individual self-determination be socially justified? With wide variations, answers still overwhelmingly lean towards Hobbes' solution wherein the 'social interest', the common interest of a society's members, can justify overriding individual rights to self-determination. The Parsonian solution, for example, seeks social order through the individuals' internalization of the common good and interest.

Marx famously claimed that the 'public interests' used to justify government policy are invariably the interests of the ruling class, fueling a debate that continues to this day. The idea that the so-called 'common good' or 'public interest' is a mythological or ideological construction that serves to conceal particular interests while excluding various groups of citizens (for example, women, minorities, workers etc) has become almost commonplace in much of the developed world. This demythologization has in many respects advanced the cause of self-determination. For example, feminism has argued that in a patriarchal society the 'public good' is in fact the men's good and women are excluded, except to the extent that they fulfill subservient roles prescribed to them in this view of society. In its struggle for liberation from traditional shackles, the women's movement has not only argued for women's rights to self-determination but also strongly encouraged women to exercise this right. To this end the movement stresses the importance of a strong 'sisterhood', or 'community of women' to provide a structural foundation from which individual women can exercise this right. Here, however, the community is reduced to an instrumental means to the ends of individual self-determination, a position akin to the utilitarianism or expressive individualism that Bellah and his co-authors critically referred to (Bellah et al. 1985). How these individually self-determining women will nurture the community and solidarity that delivered their 'independence' and 'freedom' remains an open question.

Hence we find ourselves in a situation that many social theorists refer to as a crisis in modern society. How can self-interested and self-determining individuals be motivated, attracted or oriented towards communality and active involvement in the problems of the community? How can purposive incentives and solidarity incentives be made to appeal to selfish individuals? These issues remain the central concerns of civil society. As Bellah et al. explain

> The radical individualism which tries to cut off individuals from each other actually does not make a strong individualism but a weak individualism... the individual and the community are not in a zero-sum relationship where, as one becomes stronger, the other becomes weaker. Rather, to support a certain strong individualism, a certain strong community is needed. (Bellah et al. 1985/1991: vi–vii)[4]

In other words, as previously mentioned, excessive individualism could destroy the foundations of individualism. Putnam's empirical research using a variety of data about weak social networks and individual isolation in the USA in the 1990s has confirmed that this is not merely a theoretical concern (Putnam 2000). To redress this social fragmentation Bellah advocates an 'ethical individualism deeply rooted in society' where 'individuals and the community mutually supported and enhanced each other' (Bellah et al. 1985/1991: vi–vii).[5]

The basic problem confronting public philosophy today, then, is to develop practical and effective approaches to revitalizing local communities, creating new solidarity and addressing numerous chronic social problems such as welfare, environmental conservation and gender equality. These are the challenges facing the proponents of a broad-based civil society.[6]

Regenerating the public sphere

The public sphere

Habermas' 'bourgeois public sphere' (*bürgerliche Öffentlichkeit*) refers to a critical public situated against the public authority—a general public against the State. This is a historical conception concerning the formation of public political opinion, and is generally understood to have emerged as civil society separated from the modern State. Specifically, it refers to the critical discussions of ordinary citizens in public places such as salons, cafés, or the media, and is a means of reaching social agreement in cooperation with others (Habermas 1990). The Habermasian idea has recently been translated into Japanese as *kōkyōken* (public sphere), following Hanada (1996: 24–6).[7] Replacing the dominant idea that the public is the State with a more Habermasian idea of the public sphere is therefore essential to reviving a civil society that has been subjugated to an enlarged State and mass media.

An interesting point in Habermas's analysis is that the 'debating public' engaged in face-to-face communications in the salons and cafés of early modernity was transformed into a 'reading public' through the mass production of printed media such as magazines and newspapers, and then further transformed into a 'consuming mass' through the development of new media such as radio and television. In this regard technological developments have clearly been a central factor in defining changes to the public sphere and the place of the citizen in late-modern societies. Habermas' book *The Structural*

Transformation of the Public Sphere was first published in 1962, and was not translated into English until 1989, when it received a huge response. As he acknowledges in the Preface of the second German edition in 1990, the original analysis was limited by the restrictions of the era, including the political context of the Cold War, changes in communications technology, digital media and mass media, and especially developments in media studies.

Although Habermas analyzes the public sphere as a historical concept, I define it here as a theoretical or normative idea. The 'public sphere' is then a place for the formation of public opinion and social agreement, a place where interested parties gather to debate the 'common good' or 'public interest' in free, equal and open discussions, and a place from which to critically observe and challenge the State and public authorities. Activities in the public sphere are not limited to discussions, but based upon such discussions, social practices are shaped and the values of 'publicness' and 'communality' are realized. Through these, the public sphere can be a place of political education and generational succession, in which the current generations actively work in the interests of future generations. As discussed in chapter 7, the law court is a special type of public sphere in which citizens can participate as plaintiffs or in support of plaintiffs. In principle and in effect, various media, schools, symposia and fora, and citizens' sector networks are all public spheres of one kind or another. The site of the accelerating social activities of this type are what is referred to as the 'new public sphere' in this book.

The public sphere in the information age

The information technology revolution—exemplified by digital media and the internet—has opened radically new public spaces. The development of 'media that citizens can use easily and relatively cheaply to exchange large amounts of information rapidly with various parts of the world while freely editing and processing it', has created a so-called 'network public sphere', a 'citizens' own social space independent of the economic systems and free from State control' (Hoshikawa 1994: 339–40).

The new public sphere of cyberspace is obviously not an all-purpose cure for social problems, but it has created new possibilities and means of collaboration between citizens, professionals and the mass media that may prove to be an indispensable tool for social movements, citizens' movements and NPOs. Okabe (1996; 2000) discusses examples of the financial role of intermediary organizations in the USA. Matsuura's (ed. 1999) analysis of a movement that succeeded in forcing the abandonment of plans for a landfill in the Fujimae wetlands in Nagoya City explains that it was supported through an internet mailing list.

The internet crosses institutional boundaries and reduces the relative authority of professionals, journalists, bureaucrats, and business executives, making it possible for a member of the general public to participate in public debate on a relatively equal footing. This mode of communication, in principle, tends to prioritize the content value of information over the social status of the informant.

Socially and spatially isolated individuals, however, cannot immediately develop fruitful communication links even on the internet. Although the internet is a vast and unregulated communications web, information is (and must be) organized around nodal points. The key nodal points for expanding the participation rate in this public sphere will be provided by NPO/NGOs. Social appeals and advocacy movements by NPO/NGOs are important public opinion forming activities today. In turn, the internet has produced new opportunities for recruiting more contributions to and involvement in NPO/NGO activities. The possibilities include making introductory information available to anonymous inquirers, making it easier for individuals to assess the degree of shared values and objectives with no threat of being pressured by recruiters or fundraisers, enabling on-line membership and payment arrangements, mailing list updates and interconnectivity to other organizations etc. Thus the internet can enhance communication and offer benefits in both directions, but only on the foundations of a certain level of communality.

These new public spheres remain limited in social influence if they do not receive institutional support and interconnection with the policy-making process, and remain at risk of becoming self-satisfied and self-contained peer groups. With Japan's closed and rigid public policies the need for a 'multi-layered formation' and 'diversity in opinion' in the public sphere is especially acute, as Funabashi notes (1998: 211–12).

How the public sphere will be opened for citizens to participate in the debates and decisions about public issues and to what extent information technology or NPOs can facilitate this public sphere are the central issues for advocates of a civil society in Japan today.

The public good as a normative standard for policy evaluation

Monopoly on defining the public good

In the Japanese mass media and in everyday language the word 'public' is most often used in reference to public works, especially large-scale public works. In the Nagoya bullet train and Osaka Airport lawsuits (see chapter 7) the government's definition of the 'public good'—endorsed by the court—was that the corresponding activity had to have wide social utility, and that the main actor was a 'public entity', either government or quasi-government. From this perspective 'public works' are quite simply projects carried out by the government or its agents. This interpretation of the 'public' has been used as a magic invocation to justify myriad business activities and reject or override external criticism. It has been used to justify reducing an industry's responsibility for anti-pollution measures and to avoid aid to the victims—and in many cases these justifications have been ratified by the court.

As long as the government retains a monopoly on defining the public good there is little space for a civil society in which the citizens are recognized as the legitimate bearers and shapers of public opinion, and in which the 'public good' is identified with an unspecified citizenry that sometimes stands with, sometimes against, and sometimes independently of the government. Until quite recently the Japanese national government and local governments have suppressed and obstructed the participation of citizens in public discourse and remain so inclined today—especially the national government.

In both Europe and the USA, the era of large-scale dam construction is over and serious discussions about the end of nuclear power plants have begun. More generally, the TVA set a standard and created expectations that large-scale public projects could stimulate and support economic development in remote and 'backward' districts, but concerns about their social and environmental costs have increased worldwide. Opponents argue that such projects are unnecessary, destroy both the natural environment and local communities, and incur enormous financial costs that are invariably detrimental to the government. Increasingly restrictive environmental laws and laws guaranteeing residents' participation in planning and approval processes, the expansion of environmental values, increasing difficulties in finding suitable sites and the corresponding costs are among the myriad interconnected factors that have resulted in fundamental changes in the direction of public policies, from a continuous growth perspective with its emphasis on increasing supply, to a sustainable future perspective focused on reducing the human impact on the natural environment, and therefore on controlling demand and seeking more efficient ways of using energy. This shift is increasingly apparent worldwide in water management policies and practices, as well as energy and electricity policies (see chapter 12 and Hasegawa 1996c; 1999a).

In spite of these changes, Japan's political and governmental structures have remained largely closed to the public sphere. Therefore, once a large public works project is approved, it is virtually unstoppable, even if the need for it is clearly dubious and there are valid concerns about its environmental impact. In other words, Japan is a society divided by its understanding of the public good between advocates of open public determination by the normative standards of civil society and the proponents (and practitioners) of authoritarian determination by a closed technocratic government sector. Opposing views about the validity of 'public referendums' as a means of reaching social agreement on policy directions correspond to this polarized view of the public good (see chapter 7). In a civil society, the public good is not determined by unilateral decree by a sovereign authority in order to justify its public works agenda, but must be determined through broad and democratically reached social agreement about the proposed works' social utility, the assessment of the plan's social and environmental impact, and the proper procedures for deciding (see Miyamoto 1982).

The minimum conditions for approving public works must be: a wide social utility, with no accompanying environmental destruction, and a broad social agreement. These, in turn, require that alternatives be genuinely considered, free and open disclosure of relevant information, the inclusion of all stakeholders as early as possible in the decision-making process, with appropriate means for interested parties to express their opinions while ensuring transparency, explanatory responsibility, social justice, and efficiency. If any one of these conditions is not met, the proposal is, by definition, not in the public interest.

When the normative standards of civil society are used to evaluate policy, environmental impact and strategic policy assessment procedures must be implemented at the very earliest planning stages, including evaluation of alternative plans, and once the project is operational, monitoring and redressing social and environmental impacts. After long negotiations that began in the 1970s between the Environmental Agency and MITI—with the latter enjoying strong support from industry, especially the electrical power industry, and the ruling party—a law to this effect was finally passed in Japan in 1997, becoming effective in 1998. This made Japan the last of the 29 OECD countries (at the time) to enact such legislation (Kaku 2001). But the system still requires further legislative and institutional changes to ensure the effective and total disclosure of information, genuine citizens' participation etc (Harashina ed. 2000).

Who can participate?

Who should participate in these processes? In Germany, the need to involve every concerned body at an early stage in the planning process is called the 'principle of collaboration' (*Kooperationsprinzip*), and is a basic requirement of environmental laws (Ōkubo 1997: 37). As we have seen, elsewhere, animal rights activists have claimed legal rights for non-human stakeholders in determining public policy.

Torigoe argues that simply residing in an area may entitle someone to have a say about any changes to the local environment. Using Kobe City as an example, he claims that 'based on the fact of living there...the residents in the area have a certain historically based spirit of ownership, from which they derive the right to express their judgment on environmental changes in the area'. He calls this right the 'communal seizin' (spirit), which is shared between local government and residents' groups (Torigoe 1997: 66–7).[8] This argument is notable for locating the basis of environmental rights in traditional practices. However, Torigoe also describes the 'Toga River Preservation Committee' campaign as being fought 'through the fine balance and long history of a tug-of-war' (Torigoe 1997: 79). In other words the case is illustrative of a conflict over differing interpretations of the public good rather than a general agreement about the communal seizin. This raises myriad problems. Accepting that there is in fact a right attached to a communal seizin, it is not an unambiguous and unchallengeable determinant of the public good. Where are the geographical boundaries that define which residents are entitled to claim rights on the basis of this communal seizin? Is communal seizin an individual right or a right only for a collective residents' group? These questions remain.

New public sphere: Age of the 'citizens' sector'

For a long time in Japan, as in Western societies, the responsibility for supplying public goods and services has been understood to lie with the government and private enterprise. Since the 1970s, an independent 'third sector', the 'citizens' sector',[9] has emerged and has received much attention.[10] Consisting of NPOs and social movement organizations, the citizens' sector is called the 'non-profit sector' in the USA, where it is considered to be the actual bearer of 'civil society'. In Europe it is often called the 'socio-economic sector' (Tomizawa and Kawaguchi eds, 1997).

Numerous background factors must be examined to understand the emergence of the citizens' sector at this point in history. The primary factors are typically grouped under the terms 'government failure' and 'market failure', but we must also note a 'communal failure' at the level of the family—both nuclear and extended—and the local community. Each of these terms refers to a multitude of factors.

'Government failure' refers to the failed social experiment known as the welfare state, which led governments around the world into unsustainable debt and occasionally financial crisis. The term also refers to growing doubts about the validity of a 'majority rules' democratic system from the perspective of protecting the rights of minorities. In Habermas' terms, the welfare state extensively colonized the life-world. Many have questioned the capacity of a behemoth state bureaucracy to actually deliver social welfare, citing problems such as intergenerational welfare dependency etc. At a different level again, the failure of the welfare state demonstrated the limitations of human rationality to predict and control the outcomes of its endeavors, especially in social engineering. This, in turn, raised doubts about the inherently bureaucratic structures of the modern nation-state, which, as Weber observed, only function through a 'bureaucratic dysfunction', as formal (procedural) rationalism almost invariably dominates the original substantive rationalism. This rationalism too, colonizes the life-world, making it more rigid and inefficient, as exemplified in the final decades of the Soviet Bloc's 'socialist experiment'.

'Market failure' in this context refers to the inherent contradiction in attempting to supply public goods through the market. Public goods are by definition non-exclusive, 'open to anyone', as illustrated by the police and fire services. Market sector enterprises, in contrast, are intrinsically non-inclusive—discriminating in accordance with the user-pays principle. The market thus tends to exclude the poor and less powerful members of society. It has no capacity or mechanisms to deal with problems of inequality or injustice, or so-called 'external diseconomies' such as pollution and environmental destruction. As industrialization and urbanization 'advanced', family networks and communities gradually receded, entrusting more and more of their functions to the government and market. The interpersonal relationships themselves were thereby weakened and, as mentioned above, there were clear signs of growing dependency upon the state.

The citizens' sector grew out of the vacuum caused by this triple failure, and with each new 'success' raised social expectations that it might heal some of the social malaise indicated by the terms 'government failure', 'market failure' and 'communal failure'. According to Pestoff (1998) and others, the citizens' sector's strength lies in its flexible and targeted response to the particular needs of individual lives, its experimental and alternative character and its diversity. It is especially sensitive to the needs and perspectives of society's weak and disadvantaged, and may foster a new social solidarity through voluntary participation. The citizens' sector is thus able to mediate between the government/market and family/community sectors, and initiate social reform regarding social welfare and environmental issues. One example of this is the green power schemes discussed in chapter 10, where an experimental green power program initiated by an environmental NPO was soon incorporated into government policies.

Of course the citizens' sector is not without its own problems. Salamon, a prominent researcher of NPOs, discusses a phenomenon that he calls 'volunteer failure' and points out four different types: a general shortage of necessary resources; disparities in supply and demand resulting in resources not being allocated where they are most needed; charitable paternalism; and 'amateurism', where the layperson's opinion is prioritized even when professional advice is needed (Salamon 1995: 44–8). All four of these types of volunteer failure have been repeatedly experienced in Japan's environmental and citizens' movements.

The NPO is of central importance to the citizens' sector. In short, an NPO can be considered to be an enterprise for citizens' activities. For environmental and welfare NPOs to be effective, a system is needed to ensure that the supply of human, information and capital resources is sufficient for the new network to expand. There is an important role here for specialized 'intermediary' NPOs as support centers, as intermediaries between the government/market sector and the NPOs that provide services directly to the public. The NPO law of 1998 provides a legal framework for this system.

The citizens' sector is expected to work in collaboration (as defined above) with government authorities and private companies, maintaining a close but tense relationship in order to strengthen the public's capacity to monitor and challenge the activities of the government and private sectors, to disclose social and public issues and the facts behind them, and to strengthen the mechanisms for investigating problems, proposing policies, and proposing alternatives. In short, the citizens' sector is the foundation of the new public sphere. The NPOs are clearly the most promising vehicle for enlisting broad public participation and contributions, and for delivering public goods such as advocacy, networking, social services and policy initiatives for this citizens' sector.

12 Environmental movements and the new public sphere

Towards a sustainable society

Perhaps like any historical era, it is easy to see contemporary society as being in a state of transition or at a turning point, but it is less easy to predict in which direction it is turning. That modern industrial society has reached a dead end or a crisis is widely accepted, but escape routes are not so apparent. Difficult to describe in positive terms, the present and future are commonly spoken of in negative terms such as 'post-industrial', 'post-modern', 'post-structural', 'post-colonial' and so on. The charts and compasses that modern societies created and followed towards 'progress' and 'development' have failed us and we are left standing uneasily in a fog. This, it seems, is the dominant interpretation of our times. In Japanese society, in particular, with its renowned propensity for importing Western intellectual trends, numerous insufficiently developed theories in turn become fashionable and are subjected to a vicious intellectual game of dissection and nihilistic appropriation. In this game, the loss of our bearings tends to be excessively discussed.

But in fact, modern society has not totally lost its bearings—there is no dearth of concepts or core values that can define new directions. Various Japanese notions include, *seijuku* (maturity) or *kyōsei* (conviviality),[1] while the international community speaks of 'sustainability' and 'civil society'. One of the most pressing issues in contemporary sociology, and the social sciences more generally, is clearly the need to turn its critique of societal conditions away from a disguised and futile nihilism and towards a positive engagement with prescriptive ideals for instructing and guiding the future.

Needless to say, problems abound in the use of these concepts. Politicians, government bureaucrats, media commentators and marketers, even when a precise meaning has been elucidated, can rapidly transform any such concept into a fashionable buzzword and render it vacuous through excessive and indiscriminate use. Terms such as 'sustainable' and 'convivial' are very attractive and convenient buzzwords for numerous reasons.

First, with the end of the Cold War, the ideological opposition that long defined industrial modernity has disappeared, and political adversaries are therefore difficult to identify since there is no stark basis upon which to exclude them. Conviviality is an attractive concept in this environment, for it is not possible for one's adversaries to identify themselves in opposition—there is no political advantage in identifying oneself or one's party as 'anti-convivial'. Expressing antagonism to values such as conviviality and social maturity is so taboo today that we can safely say that these concepts describe a shared social value of the post-Cold War era.

Second, conviviality and maturity do not arise from a subject-object relationship. It's not that a certain subject can make an object convivial as a characteristic of its existence. 'Convivial' is a kind of 'situational definition'. Third, again particularly in Japan, conviviality already has an important place in the traditional Japanese social imaginary where a circumstantially moral accord in group or reciprocal behavior is highly valued. It is therefore quite difficult to claim such an orientation as a 'new direction' for the implication that one (or one's party) is *only now* ascribing to this deep-seated value. In short, the term conviviality seems capable of an 'infinite embrace' and is thus an attractive choice when, for example, attempting to justify exempting various aspects of government or corporate activities from scrutiny or accountability for related social problems.

And yet, the fact of the deluge of terms such as 'conviviality', 'maturity' and 'sustainability' in vacuous discourse does not mean that we must reject them in our quest for future directions. Rather, it means that we have no choice but to embrace powerful notions with multiple-meanings. That is, in spite of their limitations, these concepts can nevertheless be used to express alternative values in a society predominantly oriented towards high economic efficiency and growth.

Before proceeding it is necessary to note that the pursuit of alternatives to the existing society is founded on an acceptance of the predictions of global warming and the evidence that such warming is a direct result of the energy consumption patterns of modern industrial societies. Furthermore, even if the values of ecological sustainability are now widely accepted, social engineering a 'soft landing' as a 'convivial society' fully transformed from a growth oriented industrial society to a mature 'growth management' type of society will demand an as yet unrealized 'self governing civilization' (Masamura 1989: 298). Environmental sociologists are therefore seeking the social systemic requirements to advance changes towards a 'sustainable future', setting aflutter the wings of the sociological imagination at the level of Mannheim's 'general knowledge' and Popper's 'technical knowledge'.

Environment and city

Japanese followers of the Chicago school of urban sociology, for example, Susumu Kurasawa (1977), continue to assume the relative self completeness of cities as a premise, emphasizing the independence of social relations and lifestyle structures in urban communities and thereby virtually ignoring ecological problems, such as environmental problems, growth management and harmony with nature. They also neglected the fact that the quality of 'the environment' codetermines—with social capital such as telecommunication and energy infrastructure—social relations and the basic amenities and quality of urban lifestyles. Of course, urban sociologists were not the only ones to have ignored these factors; since the industrial revolution, as populations and industrial activities rapidly escalated, city planners, investors and residents have increasingly divorced themselves from the self-contained environmental cycles that had been woven into the cultures of traditional rural societies. In the process cities have destroyed the eco-system, giving rise to myriad environmental problems and reducing amenity in urban environments.

Whether it is valid to claim that urban sociologists have become stuck in outdated urban paradigms or not, beginning at least ten years ago, even in Japan, urban planning professionals and local government administrators have included 'the environment' in their design codes and urban planning policies as a fundamental parameter of the highest priority. For example, many local governments in Japan have joined a municipal association that advocates the philosophies of the 'Local Government for Environmental Initiative' (Suda et al. eds, 1992: 4). The Coalition of Local Governments for Environmental Initiative Conference has been held annually since 1993. Among the 71 participants in the Conference, most are relatively small cities, towns and villages (*http://www.colgei.org/member/indes.html*, 31 October 2002). Another example can be found in Sendai. The city has adopted a 'City of Forest's Environmental Plan' that stipulates 'urban growth management' and envisages becoming a 'recycling-based city'. This future vision provides a guiding framework for general urban planning and specific building regulations etc.

The close connections between environmental policy and urban policy are already common knowledge in most of the world's major cities, which have been compelled to implement effective growth management strategies under the constant pressure of high rates of development or population increases. In the struggle for a sustainable society, there are a few sweeping generalities that can be made regarding solutions to environmental problems: it is essential that (a) the urban spatial structure, (b) local government administration, (c) industrial production activities and (d) citizens' lifestyles are changed so as to substantially reduce the environmental burden. 'All administrative areas and every local government department has come to have, more or less, some connection with the environment. Therefore, every department and section must review and improve their own ways of working from an environmental perspective' (Suda et al. eds, 1992: 30). Needless to say, this kind of self-review and self-transformation from the perspective of minimizing environmental burdens, is not only required of local government, but of industry and the citizenry alike.

In recent years, environmental movements in areas that were once seriously damaged by pollution—such as Kawasaki, Minamata, the Nishi-yodogawa district in Osaka, and the Mizushima district of Kurashiki—have instigated numerous local government and citizens' initiatives to regenerate their environment (Nagai et al. 2002). As well as changes that strengthen pollution control policies, these movements are making concrete efforts towards environmental education, recycling, tree planting and fostering various environmental enterprises, all aimed towards alleviating the downstream affects of industrial and urban pollution. Minamata, for example, infamous for being home to one of the first large-scale outbreaks of 'industrial disease' (the 'Minamata mercury poisoning disease'), has apparently 'learned its lesson' from the experience and is aiming to become a zero-garbage town in terms of both general and industrial waste. Garbage is currently divided into 21 different types before collection, making this one of Japan's most advanced garbage recycling systems. Calling for the community to 'moor again' (to reconcile and cooperate), regional resource maps and watercourse charts are being produced, and colorful community events aimed at increasing social awareness of the natural cycle of the river basin are organized and held through the cooperative efforts of the local government and citizens.[2] In short, after long years of neglect and destruction from pollution and related downstream problems, concrete methods to eliminate environmental damage are being developed and implemented.

Internationally, the International Council for Local Environmental Initiatives, with over 400 affiliated local governments from 62 countries and the EU, has become one of the pillars of environmental policy. It has been pouring resources into achieving its supreme objective—the creation of sustainable cities, traversing all geo-political regions. To this end it aspires to unify urban policy at the respective regional, national, state or prefecture, and local levels. The key issues involved in building sustainable cities are: managing growth, controlling automotive traffic, waste management, creating spaces in harmony with the environment, reducing energy consumption and adopting renewable energy resources (Samuta 2001; Morotomi 2002).

Contemporary ideological axes

Ecologism vs. industrialism

With the end of the Cold War, ideological axes have supposedly disappeared. However, it is still possible to interpret today's orientations to political, economic and social issues in the new public sphere along the axes created by opposing two pairs of values: ecologism vs. industrialism, and market vs. planned economy.

In simple terms, 'ecologism' corresponds to the value expressed in Dunlap's 'new ecological paradigm' and as such is opposed to the prevailing anthropocentrism or 'human exceptionalist paradigm' (Catton and Dunlap 1978). Ecologism stresses, first, that humans are merely one of many species in the ecosystem and is highly critical of the presumption, prevalent in the social sciences, that human beings have a privileged position in the overall scheme of things. Second, it recognizes ecological limits to human behavior, including economic development—limits determined by the rights of other living things to be free of the impacts of human society on their living environment. In contrast, industrialism rejects the romantic optimism of ecologism, stressing the independence of human action from living things and the physical environment and embracing the idea that the good society can only be achieved through unfettered economic growth and technological development.

In this context, 'sustainability' and 'sustainable development' signify an attempt to mediate between the perceived opposition of ecological and industrialist values. As is perhaps most evident in the developing countries, the meanings of such terms in actual use vary widely, tending to move along the spectrum towards industrialism when used by the advocates of economic development and towards ecologism when used by those who advocate more restrained economic growth, less vigorous consumption and more efficient use of natural energy resources.

Market vs. planned economy

Proponents of a free market economy point to the limitations of human rational-planning to satisfy the needs of selfish individuals. They argue that social goods and capital resources are most efficiently distributed through the free activities of the marketplace. From this perspective, government intervention in the market introduces unnecessary inefficiencies to the system. The most recent proponents of this small government stance are called neo-conservatives. In contrast, advocates of a planned economy prioritize social justice over economic efficiency and point to the market's incapacity to deliver social justice to explain the necessity for political intervention. Needless to say, the former Soviet Bloc 'socialists' and more moderate social democrats subscribe to variations on the planned economic system. In practice, all of the European nations have vacillated somewhere in between a planned and a market economy for many years, alternately privatizing and nationalizing various industrial sectors. If one can stand clear of any ideological bias, it seems clear that the history of the 20th century conclusively demonstrated that neither a 'totally planned' nor a 'totally free' market is either effective or desirable—or even possible. Nevertheless, this pair of opposing terms is representative of fundamentally opposing perspectives about the best mechanisms for the distribution of social and economic resources and the appropriate directions of economic policy.

If we set these two axes perpendicular to one another, with the ecologism/industrialism axis horizontal and the market/planned axis vertical, as in Figure 12.1, the crossed axes map a conceptual field of socio-economic types. In the top right hand corner, where industrialism and the market economy overlap, we find the neo-conservative paths of Reagan and Thatcher. Deregulation, restructuring and privatization are the reform prescriptions of choice for neo-conservatives. Social democracy must be placed on the planned economy side of the axis, and is divided into two different groups on the ecologism/industrialism spectrum: a strongly industrialist orientation (the bottom right hand corner) and a strongly ecological orientation (the bottom left hand corner). In Europe for example, the right faction of social democrats is industrialist, providing the main support base for the trade unions in England, Germany, Sweden and elsewhere, and is distinctive for its characteristic representation of particular interests. In contrast, the Green Party and environmental conservation movements, whose support or collaboration is crucial to the left factions of the social democrats, are ecological. The lower right hand corner of the graph represents the general orientation of the socialism of the former Soviet Bloc nations. 'Conviviality' as I defined it earlier, can be located in the lower left hand corner. Until recently ecological value and market economy values have been understood to be fundamentally opposed, with advocates of the market generally seeing ecological values as inefficient 'externalities'. Today however, it is increasingly understood that there is a place for ecologically conscious management in private enterprise (Callenbach et al. 1990), an approach that is situated in the top left hand corner where the market economy and ecologism overlap. Still in an embryonic stage, ecologically conscious management practices attempt to discover the keys to a sustainable society and environmental preservation through a new sense of enterprise responsibility and new enterprise ethics.

Figure 12.1   The contemporary ideological axis

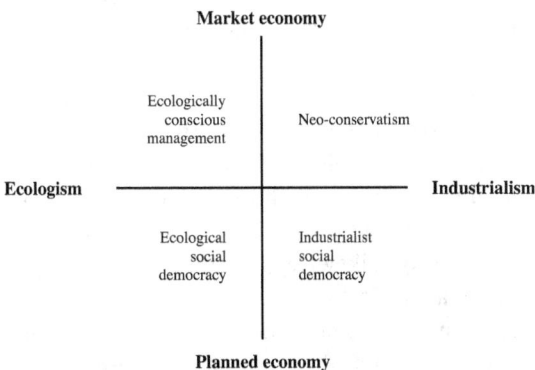

The environmental and economic policies of governments, social movements and industries can also be arranged according to the field defined in Figure 12.1. Broadly speaking, the 'Western world' sings the praises of unprecedented economic growth through a 'mixed economic system', while the other, former 'Soviet world' can be characterized by planned industrialist economy. Both orientations embraced economic policies for accelerated industrialization throughout the 1950s and 1960s. Throughout the 1980s the West became increasingly neo-conservative, a trend that accelerated in the early 1990s, after the collapse of the Soviet Bloc. A new era of enthusiastic industrialist capitalism ensued, spurred on by continuous economic deregulation, decentralization and the privatization of previously public enterprises. During this period the environmental movements assumed the role of monitoring and challenging this otherwise virtually unfettered development. These dynamics of divergence and increasing levels of antagonism between neo-conservatives and environmentalists in various European countries, the US, Canada, Japan and elsewhere, both at their respective domestic levels and at regional and global levels, have continued throughout the 1990s and into the early 21st century, redefining the lines of opposition in global politics.

Avoiding diseconomies of scale

Towards avoiding diseconomies of scale

The most apparent material manifestations of 'modernity' are exponential growth in foodstuffs, industrial goods and population. The population of Japan from the beginning of the 18th until the middle of the 19th century is estimated to have been extremely stable, increasing by only 10%, from 30 to 33 million. From 1887 to 1975, with the exception of a decline during World War II, the population grew on average at 1% per year (Kōseishō Jinkō Mondai Kenkyūjo (Ministry of Health and Welfare, National Institute of Population) eds, 1994: Tables 1–5). Clearly, 'modernity' in Japan was a period of population growth from 1887 to 1975 in comparison to the pre-modern (pre-1887) and post-modern (post-1975) periods of population stability. Likewise, per capita GNP during the modern period increased about 25 fold.

The orientation towards continuous economic growth entails the pursuit of economies of scale—a 'bigger is better' belief based on the mass-production principle that the unit cost decreases as the unit volume increases. The pursuit of this value in Japan can be seen typically in the large-scale development projects identified in the New Comprehensive National Development Plan of 1969, released at the height of the period of rapid economic growth. But the pursuit of economies of scale was soon confronted by the inherent drawbacks of unlimited growth in the form of environmental degradation and pollution problems downstream, which became all too apparent during the 1970s. Coinciding with the two 'oil crises' of 1973 and 1979, these downstream problems raised serious questions about the merits of pursuing economies of scale and unlimited material accumulation. The turn towards 'sustainable growth' indicates recognition that for growth to continue, the *dis*economies of scale must be avoided. This understanding has guided the global trend towards small-scale, decentralized, resource responsible societies. The risks referred to in Beck's conception of 'risk society' are the potentially catastrophic risks created by the unrestrained pursuit of economies of scale, or the other way around, a society that threatens to destroy itself through the diseconomies of scale.

Changes in the pursuit of economies of scale and efforts to avoid the diseconomies of scale can be observed in various social fields other than environmentalism, in restructuring measures such as computer miniaturizing, corporate downsizing and management spin-offs to form separate companies. The diseconomies of scale of the centralized planned Soviet economy finally ended with the bankruptcy of an entire social model. If we view the hyper-pursuit of economies of scale as intrinsic to the rapid economic growth from the 1950s to the mid 1970s, then the period from the mid 1970s to the mid 1980s was a time of transition as we became increasingly aware of the real potential costs of large-scale development and the shortcomings of big government. Since the late 1980s we have begun in earnest to confront the problems of reforming the system so as to avoid the diseconomies of scale and environmental risks. This reform has begun in individual enterprises as well as in national and international systems.

To illustrate these changes in slightly more concrete terms, let us now look at several social problems confronting local and regional governments. The first case concerns sewerage treatment. In many prefectures of Japan, rather than each municipality constructing their own sewerage treatment plant, a 'river basin sewerage system' was planned and built, comprising long water channels to gather the waste water from myriad distant towns and municipalities and treat it in a single large-scale sewerage plant near the sea. The project was approved on the merits of its economies of scale, with proponents claiming that it would be cheaper per unit cost to construct, operate and maintain. However, in her analysis of this project, Junko Nakanishi criticizes the scheme as having no more justification than 'advancing the self defined objective of massification' (Nakanishi 1983: 213). From her observations at the site and other available data she was able to clearly demonstrate the disincentives and diseconomies of a project of this scale, including the in fact higher cost of building such a large sewerage treatment plant, the social tensions produced in negotiating an agreement with the rightful inhabitants and surrounding residents of the designated site, and especially the incentives for the various waste-water handling enterprises involved in the scheme to cut corners, which in turn undermined the river's in-built purification process and caused the river to dry up. Nakanishi and her associates' studies of these projects have substantially contributed to changing Japan's sewerage treatment policies towards smaller, localized systems.

The second case concerns rainwater catchments. To keep-up with the long-term growth in water demand in Japan, massive dams have been built in the upper reaches of many rivers. After decades of dam-building to supply ever-increasing rates of water consumption, though, there are very few suitable sites remaining in the upper reaches for new dams. Furthermore, this huge level of urban water consumption has required the dislocation of the traditional residents of the mountainous dam sites, putting homes and villages underwater. As the discontent of dislocated residents grows, alongside environmental objections and the ever diminishing availability of suitable sites, fears of water shortages in metropolitan areas have become more acute and rising water prices are becoming significant disincentives to water consumption.

The primary cause of this water shortage is not a shortage or an insufficiency in water catchments, but in the construction practices and standards of modern cities, which have seriously damaged the traditional and natural water cycle. Because urban roadways are sealed with asphalt and concrete, rainwater no longer permeates to the underground water table. Instead it is redirected by storm water pipes to rivers and the sea, resulting in the loss of precious water resources. More concretely, the average annual rainfall in Tokyo is 1,500 ml. Before World War II, 40% of this, or around 600ml soaked into the ground. In the course of time, 300ml of that fed underground springs, which in turn can nourish urban life. At present, the volume of water that permeates underground has decreased by half, to approximately 300ml, while the other 1,200ml now flows directly to the rivers and the sea. Furthermore, as the ground dries out underneath the city centers, they become 'heat islands' and draw water up from the springs, causing ground subsidence, which causes structural damage to high-rise buildings and urban infrastructure. Other effects are that they increase both the risk of chronic water shortages during drier years and flash flooding in urban areas whenever there is a concentrated downpour.

In response to these circumstances, regional German cities and a few places in Japan (for example, Sumida Ward in Tokyo) have examined alternative city models from the perspective of autonomous water supplies that are harmonious with the environment. The basic components of such a scheme must include, first, that rainwater is stored by office buildings, factories and domestic residences, both for miscellaneous applications and for use in disasters. Secondly, surplus rainwater must be able to saturate the ground. Positive measures here include using permeable road surfaces and storm water systems that restrain rainwater run-off. As well as ensuring the water supply, these measures have the added effect of overcoming urban dryness and the 'heat island' effect.

The urban waste disposal problem closely parallels the urban water supply problem. Societies that consume large quantities of goods also discharge large quantities of waste. Tokyo, like many of the world's major cities, faced a crisis when it exhausted the potential sites for dumping its garbage. Garbage is a major issue confronting every local government, both large and small. Going beyond the introduction of rules for garbage and recycling collection to be followed by town and city residents, concrete efforts toward garbage reduction are now being introduced by local governments around Japan and the developed world, including simplification of packaging, more thorough separation for recycling, collecting and composting natural waste, and recycling a variety of materials. But much more is still called for. A revolution in consciousness and substantial changes in lifestyles are essential for primary producers, manufacturers, merchants and consumers, from seeing garbage as a problem for the governing authorities to deal with after the fact, to seeing the need to resolve it throughout the product cycle, beginning at the product development stage, and through the marketing/packaging considerations, and the distribution and retail processes to the final consumer. In short, every person who makes a decision or executes one from the product's first conception until its disposal must do so with an eye towards garbage reduction and recycling, in every workplace and every household. The long history of collective and systemic selfishness, of 'free riding' producers and consumers who avoid small costs by passing the waste disposal problem downstream, until it is eventually born by local governments, has burdened these governments with immense financial, social and environmental costs. If that were not bad enough, we have virtually exhausted the suitable sites for dumping all of our garbage.

Some advocates of the 'urban lifestyles', a strong theme among the followers of the Chicago school of urban sociology in Japan, have stressed the intrinsic role of specialist management systems in urban society (Kurasawa 1977). But this idea that urban life can only be sustained through specialist management is akin to the industrialists' view of scientific control, taking responsibility for trying to change water and garbage practices out of the hands of general households. They missed the role and social meaning of ecological communal-village lifestyle and the possibility of citizens' initiatives towards a more sustainable city.

Cases to redress the diseconomies of scale have the following points in common.

1. There were a variety of responses to actual regional circumstances and idiosyncrasies rather than standardized solutions imposed by a strong and highly centralized government.
2. Solutions reflected the ability of local government employees and residents to handle local problems without being wholly dependent on specialists.
3. With independent safeguarding of wards and independent ward management as a basic principle, marginalized regions did not have undesirable facilities imposed upon them by the central government, nor were they subjected to the downstream affects of these facilities.

From nuclear power to diversity in power

Nuclear power generation epitomizes the modern technological pursuit of economies of scale. Widely promoted by its advertisement 'too cheap to meter' in the late 1960s and early 1970s, governments and private utilities around the world rushed to join the 'nuclear age'. Nuclear energy's particular merit is that it can produce massive amounts of energy from a tiny amount of concentrated uranium. Industrialization is typically defined in terms such as 'the extensive use of inanimate sources of power for economic production' (Moore 1963). We have come to regard the rate of consumption of non-organic energy as an important barometer for economic development. We have also seen the long-term development of a high energy consuming lifestyle. The underlying premises for adopting nuclear energy were to maximize the economies of scale in electrical power generation while maintaining strong economic growth in a socio-economic system increasingly dependent upon high volume energy consumption.

In Japan today, the Ministry of Economy, Trade and Industry and the electric companies continue to proclaim the superiority of nuclear power, even as numerous and powerful disincentives are increasingly being acknowledged by industry and government leaders around the world. Disincentives include:

1. The huge amount of capital required to build a nuclear power plant leaves the power company with high levels of long-term debt, which has come to be seen as too risky in an increasingly dynamic global market.
2. Decreasing oil prices and a slump in uranium prices[3] reduces the economic advantages of nuclear power.
3. Technical difficulties in safely disposing of radioactive waste and the social conflicts that therefore accompany every attempt to procure a suitable new disposal site.
4. The long half-life of radioactive materials, some more than 10 thousand years.
5. The risk of a severe nuclear accident.
6. The effects of low level radioactive emissions to the environment during normal plant operations.
7. The risks of nuclear weapons proliferation.

Since 1989, coinciding with the end of the Cold War, the electrical supply structures of countries around the world have undergone major transfigurations. Typically, this has involved large national-ized power monopolies controlling an integrated power generating and distribution system—building large output power plants to ensure supply to distant consumers while promoting heavy energy use—being broken up into smaller decentralized companies, sometimes privatized, to operate in a deregulated (relatively) open-market. Although this energy market still tends to be dominated by large corporations—now predominantly multinational corporations rather than state-owned enterprises—there has also been a profusion of diverse smaller enterprises pursuing alternative energy sources, such as cogeneration, wind power or solar power. These new enterprises are much more responsive to local market needs than their behemoth predecessors, and have vigorously pursued the development of new low pollution or zero pollution renewable energy resources. They have also made rapid progress toward greater energy efficiency and energy saving measures, as well as encouraging more entrepreneurs to enter the electricity industry. I have discussed the details of this process in a case study of the Sacramento Municipal Utility District's (SMUD) forced closure of its nuclear energy facility and the successful remodeling of itself as the world's first large-scale alternative energy producer to prioritize energy efficiency and renewable energy resources (see Hasegawa 1996c).

With ever-increasing public opposition to any new nuclear, fossil fuel or hydro-electric power plant proposal ensuring long drawn out legal battles and escalating costs, since the 1990s 'least cost planning' and 'demand side management' (DSM) have become mainstream concepts in the electricity production world. These terms originate in the ideas of 'soft energy paths' advocate, Amory B. Lovins. According to Lovins' model, electricity needs can be vastly reduced by:

1. Rejecting the premise that perpetually increasing energy demand must be met—and therefore additional generating facilities must be built to meet it (and the bigger the better)— and adopting new premises that see energy as a limited resource to be used sparingly and efficiently—and therefore rising demand must be dealt with on a case-by-case basis.

2. Rejecting hard energy paths that seek to fill the gap between the projected demand and the current supply capacity with additional nuclear power and fossil fuel plants, and instead make renewable resources such as wind and solar power the primary energy sources, controlling demand by increasing energy efficiency.

3. Limiting the use of particularly high quality and high priced electrical electricity to indispensable items that have no substitute. For example electricity used for space heating and similar is inefficient. (Lovins 1977)

The guiding objective of the 'soft energy path' and the spin-off DSM ideology is avoiding the diseconomies of scale. Learning how to develop and follow these soft energy paths is fundamental for the pursuit of a sustainable society.[4]

Policy-oriented social movements

Closed structures of political opportunity

The Japanese public, urban and environmental policy-making processes are rife with structural problems. First, the national government's technocracy is an extremely centralized, planning and decision-making system, controlling approval and authorization of government revenue and expenditure, and thus wielding great institutional and non-institutional power. It is a highly efficient system maintained through a rigid hierarchy that enables central bureaucracy directives to be promptly carried through by individual local governments. This type of efficiency, however, comes with many costs and social barriers, such as:

1. A closed and secretive planning and decision-making process.
2. Homogeneous policies that disregard local and particular circumstances.
3. Rigid long-term planning that precludes adapting policies to suit changing socio-economic and environmental circumstances.
4. Local and regional government authority is curtailed, rendering them little more than 'subsidy' and 'petition administrations'.
5. Deeply entrenched power struggles between government departments, agencies and ministries over jurisdiction, authority, control etc.

A blanket decentralization law was introduced and, from April 2000 under the advertisement of 'structural reform', formally at least, an 'equal and collaborative relationship' began between the national and local governments. It is as yet much too early to reach final conclusions about the extent to which decentralization has genuinely been achieved. We shall have to wait and see.

Second, this powerful centralized planning and decision-making bureaucracy has been 'overseen' by a political system in which there was no change of government for more than 50 years (except for 11 months from July 1993 when opposing parties formed a coalition government). Under these circumstance, the Diet's performance as a critical monitor of the executive branch of government has been less than adequate. Thus the bureaucracy has been able to carry out the executive's political and economic agenda relatively unchecked by 'countervailing power' (Galbraith 1952) or democratic processes. As a result, the central bureaucracy has exercised a monopoly on definitions of the 'public interest', and planning, decision-making and policy implementation measures have all tended to ignore so-called 'external diseconomies' and the rights of minorities and other marginalized people downstream.

Local and regional governments had very little influence in planning and decision-making processes, particularly in the cases of giant national projects. Hence, opportunities for ordinary residents to participate in the process were even further limited. Very little effort was invested in attempts to learn what the general public wanted or thought about the government's projects, and thus the social expense, the full extent of negative flow-on effects, and the full scale of the affected area were easily underestimated. After years of not being heard, citizens' groups began to organize protest rallies and petitions to express their opposition to the bureaucrats' plans and modes of operating. Until the early 1990s, as these movements grew, they became increasingly confrontational, eventually forcing various plans to be withdrawn and revised.

The institutional exclusion of local governments and residents from the planning and decision-making process has produced a one-sided, state planning authority whose responses to opposition varies between strong-arm tactics and placation strategies, making a negotiated agreement between the proponents and opponents of a proposal very difficult, if not impossible, to achieve. This closed and rigid system is the primary reason that major national projects such as bullet trains, airports, nuclear power facilities and hydroelectric dams have invariably become bogged down in serious long-term conflicts. The mutual distrust between the opposing sides in such disputes often extend into the local community, further entrenching the resulting social disadvantages.

In the closed technocratic system, social movements objecting to large scale projects tended to be oppositional protest movements, and had difficulty proposing constructive and workable alternatives. Environmental movements in Japan have generally had great difficulty in casting aside these traits. In contrast, the new public sphere demands the creation of settings and systems where common starting points for rational debates can be found and the formation of agreement can gradually be achieved among people with diverse interests and values. Ideally, the new public sphere will be an integral part of a new open planning and decision-making process and lay the foundations for a long term secondary refinement of critical social movements. There is an important role in this process for professionals who remain largely independent of the government and industry sectors in carrying forward this transition from a closed technocratic system to an open democratic system. Professionals such as natural and social scientists, lawyers, journalists and other specialists are central to American and European social movements, acting as 'counter-specialists' who critically challenge the expertise of government technocrats. They also serve as channels for public opinion, translating 'the thoughts, desires and demands of residents into practicable urban planning policies and implementation strategies, while at the same time performing the role of translating specialist terminology and systems into ordinary language that can be understood by the general populace' (Öno and Habe 1992: 227). This is but one example of the criticism, exchange and collaboration required of the three groups that constitute the citizenry—local government employees, independent professionals and inhabitants— if we are to achieve a mature and sustainable society through the creation of a new public sphere.

The political function of university towns

Small university towns in Germany and America serve a similar political function to the professionals discussed above, albeit on a different scale. Where a prominent university or research institute is located in a town with a small population, the academic teaching staff, students, researchers and others associated with the university constitute a relatively large proportion of local voters and thus, through local elections, social movements and NGOs can have a substantial influence on regional politics and the broader society. In these circumstances ecologically oriented political factions have grown larger and stronger, enabling idealistic and progressive environmentalist administrations to become well-established. Freiburg, Germany and Davis, California are leading examples of this tendency.

Davis, California (pop. 46,000) is the home of the Agricultural Studies and Environmental Studies Colleges of the University of California. Davis is renowned for its advanced environmental regulations and initiatives, and is a world leader in the introduction of urban growth management policies (1968) and recycling policies (1972). The city adopted garbage separation and collection practices from 1974 and in 1975 it enacted the first energy saving ordinances in the USA. Today it is also known for its long serving Mayor Mike Corbett, who was America's foremost advocate of sustainable cities, 'ecological new towns' and 'green development', and one of the originators of what is called environmentally responsive development. 'Village Homes', a 70-acre residential community completed in 1975 is considered to be the earliest model of an ecological community or open garden city, but the underlying principles—such as (a) restricting the use of automobiles in the city, (b) maintaining a green-belt for public use, (c) wide-spread use of solar heating and hot water systems, (d) and restricting the number of residences—are becoming standard features of new urban experiments (see *http://www.city.davis.ca.us/pb/cultural/30years/ chapt04.cfm*, 10 November 2003). In the 1990s, Davis initiated America's largest experimental solar power facility, named PVUSA, which is now owned and operated by SMUD.

International opinion polls rate these German citizens as the world's most enthusiastic about environmental policies. Germany has been the world-leader in reducing carbon dioxide emissions, pledging a Kyoto protocol reduction in green house gases of 21% (1990 levels) and by 2000 had already reduced their total emissions by 19%. Meanwhile, many developed countries such as the USA and Australia remained reluctant to commit to any reductions, and others, such as Japan, continued to increase their emissions in spite of having set targets for reductions. Japan's production of green house gases had increased almost 10% by 2001 (from 1990 levels) in spite of setting a target for a 6% reduction. Importantly, in the thoroughly decentralized German government system, leadership in environmental policy comes from the local government level via the citizens' sector. Between 1989 and 1998, prompted by the growing environmental awareness among the general populace following an educational campaign by the civil organization 'German Environmental Aid Fund', a national competition was launched between local governments for recognition as the 'Federal Capital of Environmental Conservation', nicknamed 'eco-capital'.

The university town of Freiburg (pop. 200,000) won this honor in 1992 (Imaizumi 2001). Freiburg is a beautiful city with a 900 year history at the south western extreme of Germany, backing onto the Black Forest with an earthen embankment surrounding the city center. Through the initiatives of former Mayor Rolf Boehme, including progressive environmental policies and notable achievements in garbage reduction, controlling automobile traffic and reducing electricity consumption, Freiburg has become a model for sustainable living initiatives.

Having first visited Freiburg in July 1994, I revisited in March 2002 and was struck by the steady progress that had been achieved in the interim. A new train station had been built incorporating 40kW of solar power generating panels. There were a lot of new solar energy facilities for research and commercial use, including a solar cell manufacturing plant, an urban redevelopment project in the Vauban district that included a solar parking garage, and a low energy housing estate in the Rieselfeld district. The contrast to the early 1990s, when alternative urban developments remained individual and experimental, to this wide-spread construction of alternatively powered buildings and public facilities carried out through community level initiatives was breathtaking.

The environmental progressiveness of Freiburg's city administration was born during a successful movement opposing the construction of a nuclear power facility at Wyrl, 25km away, in 1975. This protest action attracted young environmentalists and student activists from all over Germany. With this huge groundswell of popular support, it became the first social movement in Germany to successfully force the cancellation of a proposed nuclear facility. The success attracted additional environmental conservation groups from all over the nation to Freiburg, setting the stage for the establishment of the largest and the most influential independent research facility in Germany—the 'Eco Institute'. This successful campaign was also a precursor to the creation of the Green Party in 1980. Freiburg has continued to be a leading stronghold of the German Greens ever since, and is a proving ground for the party's environmental policies.

As we can see then, small university towns can make excellent proving grounds for environmental groups and environmental policies, by concentrating socially aware and politically active professionals and specialists in a small area where they constitute a significant proportion of the populace. As the Davis and Freiburg examples clearly demonstrate, such environments can become hotbeds of environmentally related research and business ventures.

In Japan, where there is no tradition of this kind of university town, similar progressive/ experimental communities might instead be created in regional centers and the outskirts of major metropolitan areas—that is, in communities where the social differences between the three core groups mentioned above—local government employees, independent professionals and inhabitants–are relatively small. Ideally these communities would be located in a prefectural capital with a population of 2–300,000, a university, a newspaper, its own television and radio stations and some number of lawyers.

The function of policy-oriented NGOs and think tanks are of great importance in cultivating skills and knowledge. American and German environmental organizations such as the Lovins-led Rocky Mountains Institute, the Natural Resources Defense Council (NRDC) and the Eco Institute etc have played significant roles in changing energy and environmental policies both in their respective countries and, as a result of their local successes, more broadly. Notably, most of the wind or solar power generating facilities that have been built in these two countries have been built by private business enterprises. In other words, we must recognize that both citizens' sector NGOs, and private sector enterprises and professional consulting firms have become important resources for training ambitious, individual talent and creating an environment conducive to diverse and dynamic initiatives.

The general characteristics of the transfiguration of social movements through the 1990s can be seen in Table 12.1, which contrasts typical characteristics of the first and second stage American movements against nuclear energy. The table reveals a shift from an oppositional first stage movement, protesting and petitioning from outside of the system, to second stage movements deeply involved in the planning and decision-making processes and able to stimulate institutional reforms to government and corporate bodies. Similar changes can also be found in environmental movements more generally, as well as the feminist movement and other unrelated social movements. In contrast to American social movements, until the mid 1980s, Europe's new social movements remained uncompromisingly opposed to the idea of any type of collaboration with government authorities (which were generally equally as uncompromising towards the movement) (Offe 1985). However, since the end of the 1980s, concurrent with the end of the Cold War, collaboration between government institutions and alternative think tanks has become increasingly common, especially at the regional and local government levels.

Since the mid 1980s, offshoots from the Freiburg anti-nuclear movement, like the Eco Institute, have strengthened their second stage characteristics, with a growing proportion of their work commissioned by government agencies, private enterprise and even the nuclear power industry. The influence of the organization is evident in the major role it played in bringing about the German 'consensus agreement' of denuclearization and the appointment of Dr.

Table 12.1: Two phases of the nuclear energy problem in the USA

	First stage (Anti-nuclear phase)	Second stage (Denuclearization phase)
Time period	1970s to late 1980s	Late 1980s and 1990s
Movement objectives	Preventing the construction of nuclear energy facilities and obstructing operations.	Increasing energy efficiency. Increasing the use of renewable energy sources.
Strategies and tactics	Confronting the nuclear power industry and electric companies. Exposing hidden agendas of nuclear facilities and close relationships between the nuclear industry and electric companies. Criticizing, staging demonstrations and non-violent protests (sit-ins and the like) and filing lawsuits.	Civic participation and control of the management of the electricity supply system, least cost planning, regulatory reforms, collaboration, economically appropriate usage and normalization of electric company management.
Contentious issues	Safety of nuclear energy facilities, disposal of radioactive waste.	Diseconomies of nuclear energy, nuclear waste management problems, closure of dilapidated facilities.
Movement consequences	Estrangement from the greater populace, isolation of the movement, loss of influence and engrossment in public concerns.	High level of financial risk for investors in nuclear energy.
Values	Affinity with a protest culture lifestyle, orientation towards a simpler lifestyle, criticism of industrial civilization, distrust of the market.	Multifaceted support of new energy policy and efficiency, improved regulations and their philosophies, maintaining current standards of living and developing manufacturing standards for a market economy that is compatible with denuclearization.
Representative examples	Battles to prevent the establishment of nuclear power stations in Diablo Canyon, California and Seabrook, New Hampshire.	Reconstruction of SMUD following shutdown of nuclear operations. Practical activities of A. Lovins, NRDC and UCS.

Michael Sailer as the Chairperson of the Reactor Safety Commission of Germany in March 2002. The RSC is roughly equivalent to the American NRC (Nuclear Regulatory Commission) or the Nuclear Safety Commission in Japan. Sailer was one of the co-founders and remains one of the leaders of the Eco Institute. The nomination of this well known activist with a long history of criticizing the nuclear industry and promoting nuclear policy was symbolic of the changes in German nuclear policy since the June 2000 agreement.[5] In my view it is also symbolic of the second stage movement, a shift towards deep involvement within rather than criticism from without.

We see the change in older environmental NGOs as well. For example, compared to the World Wildlife Fund's emphasis on fostering relationships with the business and industrial sectors, Greenpeace's emphasis on direct action has strong first stage characteristics. More recently, however, it has begun to assume second stage characteristics, demonstrating its abilities in policy-making and analysis. In Germany, it began managing the nationwide consumer cooperative, Greenpeace Energy EG, in 1998 under the slogan, 'Say 'NO' to nuclear power: switch to Greenpeace Energy'. It currently has about 20,000 private customers and more than 600 business customers (see *http://www.greenpeace-energy.de/*, 1 November 2003).

As resolving the obstacles to convivial and sustainable development are increasingly recognized as public policy priorities, the types of severe public confrontations that once occurred over the construction of nuclear energy facilities are becoming less common, at least in environment related areas, and this trend seems likely to continue in the foreseeable future.

The 'new social movements' identified by Touraine were defined in terms of social movements pitted against technocrats and have similar characteristics to the first stage anti-nuclear movements. In the second stage, however, diverse collaborations between technocrats and social movements are becoming the motive force behind reforms to the socio-economic system as we move towards a mature convivial society. On one hand this means institutionalizing the movements, which could repress spontaneous energy and initiative, removing the critical tension that comes with opposition to the political authorities, and indirectly stifling critical public discourse. On the other hand, second stage, policy-oriented, social and environmental movements appear to offer the greatest potential for social reform in the post-Cold War era.

13 Transforming the citizens' sector

The citizens' sector in the 1990s

The 1990s in Japan have often been referred to as a 'lost decade' in reference to widespread political crises and the end of Japan's so-called 'economic miracle'. But the picture looks quite different when viewed through the lens of citizens' activities and social movements. Even when compared to the early 1970s, when local residents' movements around the country advanced pollution and development issues, and the development of the 'new social movements' in the 1980s and their corresponding networking methods, it is safe to say that the transformations in the citizens' sector and social movements in the 1990s were quite remarkable. The term 'citizens' sector' here refers to the *public* activities of private citizens, NPOs and social movements.[1] The citizens' sector is distinct from the government sector—including national, regional and local level legislative, administrative and judicial entities—and the commercial or 'private' sector, comprising privately owned business enterprises.

The major transformations in the Japanese citizens' sector in the 1990s can be arranged into five points. First was the rapid progress in organizing citizens' activities, following the March 1998 enactment of the Law Concerning the Promotion of Specific Non Profit Organization Activities (the NPO law) with around 16,000 groups submitting applications nationally by the end of December 2003, and close to 15,000 groups receiving corporate status.

Second, by taking advantage of local government procedures for information disclosure, citizens' movements achieved regional financial administrative reform. Third, demanding public referendums became a standard movement strategy (see chapter 9). Public referendums were held in several regions where public opinion was divided and the majority of the regional assemblies approved ordinances to enable a referendum. Nevertheless, in many cases the assembly did not pass the ordinance (as will be discussed shortly) and the movement subsequently abandoned the campaign for a referendum. Meanwhile the movements set their sights on electing their own representatives into local and regional governing councils.

This trend had clearly begun in the 1980s and steadily increased throughout Japan in the 1990s.

Fourth, although confrontational movements opposing government and private sector initiatives remain prominent, other movements increasingly became oriented towards proposing alternative policies, developed in collaboration with independent specialists, and cooperating with politicians, administrators and private enterprises to bring these policies to fruition. Fifth, citizens' activities and social movements became better informed and organized, developing strong working relationships through international networks of social movements. As international information exchange through the internet became commonplace, local objectives and strategies could be devised with up-to-date knowledge of international activities. All of these factors strengthened the policy-orientation and institutionalization of social movements.

Each of these transformations was directly or indirectly related to the steady institutionalization of citizens' activities and social movements.[2] 'Institutionalization' here means: 1. As formal entities with substantial influence in relation to specific issues, citizens' activities and social movements have been recognized and accepted by governments and bureaucrats as important and indispensable partners in developing social policy and assessing the social impact of planning and development proposals. 2. The citizens' sector and social movements have moved away from protest demonstrations and direct action, instead becoming directly involved in using and reforming the current social institutions and systems for enacting ordinances, as well as petitions, deliberation in legislative assemblies, elections etc. In short, movement activities that work with the system have become much more prevalent.

Conventionally, interpretations of the relationships between social movements and macro level social change—such as industrialization, urbanization and the aging population—have understood macro social changes to give rise to specific social problems and to be the causal conditions for the creation of responding social movements. A classic example of this is the mass society theory's interpretation of social movements in terms of 'collective behavior' (Hasegawa 1985a). As described above, when attempting to understand social movements in the context of the information age, internationalization and institutionalization, it is important to recognize that these macro level structural changes in society largely determine the social and political environment in which the citizens' sector exists, and thus the strategies required to achieve a movement's objectives.[3] Macro structural changes can then be understood as factors determining the structures of political opportunity, strategies, and the resource mobilization capacities of social movements. Each of these transformations will now be discussed in more detail.

Progress in NPO organization: Background to the rapid proliferation of NPOs

Greenpeace has offices in 27 countries, including developing countries, yet the Japanese branch was the last to be legally incorporated as a specifically non profit corporate entity (NPO corporate body) on 11 January 2002. This illustrates how slow Japan has been to reform its institutional frameworks, in spite of being an economic power with the world's second highest GDP.

While expressions of the need to recognize citizens' groups as corporate entities began to arise in the early 1990s, the Great Hanshin-Awaji earthquake of 1995 brought this to a head, highlighting the essential social functions of citizens' activities. In March 1998 the NPO law was passed by the Diet, becoming effective in December 1998, enabling the legal incorporation of citizens' sector organizations. Until then, volunteer and citizens' groups, could not even lease premises for their operations in the group's name. Subsequently they have been recognized as legal entities that are legally bound by contracts etc.

There has been an explosion in the creation and incorporation of NPOs since the law was enacted, far exceeding the expectations of all involved. This growth is charted in Figure 13.1. More than 1,000 organizations were incorporated in the first year after the law was enacted, surpassing 3,000 in the second year, over 5,000 in the third year, and close to 10,000 in the forth year. According to the Cabinet Office's data at the end of December 2003 (*http://www.npo-homepage. go.jp/data/pref.html*, 16 January 2004), 16,353 applications for NPO status were lodged over the five year period, and 14,657 applications were granted (only 55 applications were rejected and others were reviewed). In the six months from June to December 2003, an average of 460 groups were certified each month, the highest rate yet—1.4 times greater than the average for the same period in 2002 (326 per month).

*Figure 13.1   Number of specified nonprofit corporations (Dec. 1998
to Dec. 2003)*

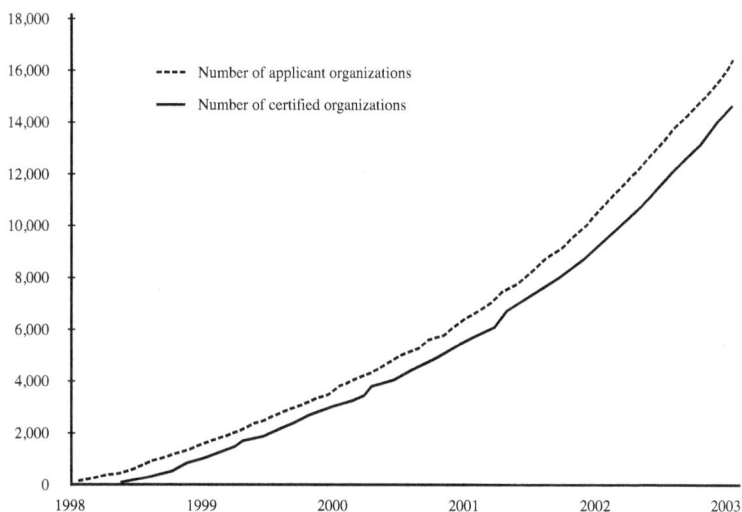

Note: Graph compiled from Cabinet Office data current at the end of December 2003.
Data were collected in December each year.

NPOs form a new 'social culture'

Social background and expectations

The Hanshin earthquake was undoubtedly a significant factor in bringing the NPO Law into effect, but it was only one such factor. The necessity for this legislation had been recognized much earlier and substantial progress had already been made towards it before the earthquake (Matsubara 1999). The structural factors behind the growing demand to incorporate citizens' groups include a diverse range of international and domestic phenomena.

Internationally, the growing expectations for NPOs are generally attributed to 'market failure' and 'government failure' (Pestoff 1998). To these we can add 'communal failure'—that is, the deterioration of the traditional institutions of the nuclear and extended family and local community bonds and their corresponding social functions. Throughout the developed countries, governments, markets, families and communities are failing to deliver social welfare and cohesion. Citizens' sector activities and organizations are therefore increasingly expected to compensate for this functional vacuum in social institutions. In Japan, the 'failures' of the government and market to deliver social welfare became acute in the early to mid 1990s, with the collapse of the bubble economy and the Hanshin earthquake. The myriad phenomena collected under the label 'globalization', especially since the 1980s, have undermined the sovereignty of national governments—both in terms of their capacities to manage their economies and markets and to defend their citizens and territory from 'foreign' threats—for example, radioactive contamination and acid rain. Awareness of the world as a single 'global commons' is growing in every society, as concerns about global warming continue to magnify. International environmental NGOs, such as the Climate Action Network (CAN), are having a major impact on the direction of international global warming conferences. Meanwhile, international environmental NGOs like the World Wildlife Fund, Friends of the Earth and Greenpeace are becoming increasingly involved in matters of governance, stimulating expectations around the world that NPOs might complement or substitute for the failed government and market institutions.

In the relatively unique Japanese context, concepts related to the 'public' and the 'public sphere' have begun to be redefined (see chapter 11). Conceptions of 'public' in Japan traditionally had strong connotations of the 'nation' or 'state'. The English conceptions of the 'public', referring to the interests and perspectives of the general populace, have had very little currency in Japan. The definition and protection of public interests in Japan has long been considered to be the exclusive prerogative and responsibility of the centralized government. Unsatisfied with this situation, and armed with up-to-date information about citizens' sector activities elsewhere, Japanese citizens began to form groups to address each and every social issue of concern, claiming the right to define the 'public interest' themselves, and began to test this 'right' through a variety of practices and avenues. In this environment, NPOs have come to be seen as the definers and defenders of a 'new publicness' and a new 'public sphere' (Hasegawa 2002b; Yoshihara 2002).

NPO issues

As this trend towards citizen-based self-help and mutual support strengthens, so too do issues concerning the availability and distribution of necessary resources for the relevant actors in the public sphere. Japanese NPOs are confronted with problems of insufficient revenue and human resources, especially full-time administrative staff. In an effort to redress this, the NPO law was revised two years after its introduction to make provisions for preferential taxation status for NPOs. From 1 October 2001, individual or corporate donations to NPOs that satisfy certain requirements were exempted from death duties, income and corporate taxes. But requirements for qualification were too restrictive–only ten applicants qualified in the first year and eight in the second year.

The shortage of human resources means that there is a tendency for a few competent individuals to be actively involved in more than one organization. Concerned parties are hopeful that experienced and highly educated individuals, including housewives who do not work full time, as well as retired private sector office workers and middle managers, government officers, teachers and union officials will fill this gap in human resources as volunteers.

The system introduced by the NPO law enabled local incorporation through prefectural governments, but it also provided a mechanism for a single nation-wide registration through the Cabinet Office (formerly the Economic Planning Agency) when an organization has offices across a number of jurisdictions. At the time of writing over 90% of NPOs are incorporated at the prefectural level, and only 9.5% through the Cabinet Office. Hence, although this system clearly encourages groups to be active in their local regions, it also hinders the development of a nationwide NPO.

In contrast to the highly pragmatic NPOs and social movements in the USA, Japanese social and environmental movements have until recently shown pronounced idealistic and ascetic tendencies. This has in part been a result of their poor relationships with government and business. As Japanese NPOs and social movements become increasingly oriented towards collaborations with government and business enterprises as practical approaches to increasing their political influence, the dangers of friction and discord with movements or factions that uncompromisingly retain their idealistic principles will continue to increase.

At the same time, these 'idealistic' concerns are well founded— the transformation of NPOs and citizens' movements into genial collaborators carries great risks for the organizations and movements. Overemphasizing cooperation and collaboration may stifle internal dissent and ostracize external critics. There is a further risk that activities will stagnate and the original objectives will be forgotten as the interpersonal relationships between members and the survival of the group become self-serving.

In other words, as increasing numbers of NPOs actively seek projects commissioned by the government, there is a strong risk that they will be effectively reduced to government contractors. To avoid this type of 'citizens' sector failure' and ensure equal relationships between the government and NPOs it will be necessary to ensure that all commissions and contracts are awarded via impartial and transparent procedures.

Another problem confronting the Japanese citizens' sector is the regional maldistribution of NPO activities. Tables 13.1 and 13.2 are drawn from the previously mentioned Cabinet Office website, listing the top and bottom 15 prefectures and municipalities ranked by NPOs per 100,000 population (at the end of October 2002). The municipalities and prefectures with the lowest ratios of NPOs/ population were generally located in isolated or marginalized areas near the Sea of Japan or in the southwest of the country. In contrast, the top 15—led by Tokyo—are concentrated in regions with major metropolitan centers. The presence of prefectures like Gunma, Mie and Shizuoka in the top 15 are notable, as they indicate the effectiveness of local government efforts to promote NPO activities. Fukui Prefecture's place in this ranking has recently jumped dramatically, following a sharp rise in incorporations. In this case

Table 13.1 NPO concentration: Top 15 prefectures

Rank	Prefecture	No. of NPOs	NPOs per 100 000 pop.
1	Metropolis Tokyo	1877	15.559
2	Kyoto Prefecture	244	9.228
3	Gunma Prefecture	178	8.790
4	Osaka Prefecture	682	7.746
5	Mie Prefecture	138	7.431
6	Fukui Prefecture	60	7.238
7	Kōchi Prefecture	55	6.757
8	Nagano Prefecture	141	6.366
9	Kanagawa Prefecture	514	6.054
10	Hokkaido Island	344	6.053
11	Okinawa Prefecture	78	5.918
12	Tochigi Prefecture	116	5.786
13	Yamanashi Prefecture	51	5.743
14	Miyagi Prefecture	135	5.708
15	Shizuoka Prefecture	215	5.707

Table 13.2 NPO concentration: Bottom 15 prefectures

Rank	Prefecture	No. of NPOs	NPOs per 100 000 pop.
1	Kagoshima Prefecture	46	2.576
2	Saitama Prefecture	198	2.854
3	Aomori Prefecture	44	2.981
4	Toyama Prefecture	35	3.122
5	Aichi Prefecture	226	3.209
6	Ibaraki Prefecture	111	3.717
7	Nara Prefecture	56	3.881
8	Akita Prefecture	47	3.953
9	Shimane Prefecture	31	4.068
10	Gifu Prefecture	86	4.080
11	Miyazaki Prefecture	48	4.103
12	Niigata Prefecture	102	4.120
13	Fukushima Prefecture	91	4.278
14	Nagasaki Prefecture	66	4.351
15	Tokushima Prefecture	36	4.369

Note: Tables 13.1 and 13.2 compiled from the Cabinet Office data current at the end of October 2002. Population figures are from the 2000 Japanese national census.

too, political efforts are succeeding in increasing the numbers of NPOs in the region.

In the past, public goods and services in Japan have, in effect, been supplied by the central government in accordance with universal criteria that often favored disadvantaged remote and marginal communities. As the rankings in Tables 13.1 and 13.2 demonstrate, however, there is now a danger that uneven rates of NPO activity between major metropolitan and remote rural areas may cause an even further widening of the disparities in the quality and availability of public services as the citizens' sector is increasingly relied upon.

Ombudsmen and procedures for information disclosure

One of the most significant developments in the Japanese public sphere in the 1990s, in terms of instigating substantial changes and improvements to local governments, was the creation of the office of the citizens' ombudsmen. Citizens' movements, through their professional members such as lawyers, tax accountants and others, utilized new freedom of information procedures, citizens' petitions (for public audits) and litigation against local governments to expose illegal expenditures, including false travel expenses, false disbursements, padded accounts, entertainment for bureaucrats, unnecessary expenditure on public projects and secretive subsidies (or 'coin tossing') to affiliated organizations by local and prefectural governments and administrations around the country. In the process, information disclosure procedures were further refined and entrenched, some ordinances were amended and wholesale reforms to public accountability procedures advanced.

Following the establishment of offices of the citizen's ombudsman in the cities of Kyoto and Osaka, the Sendai Citizens' Ombudsman's office was established in 1993. The establishment of this office was primarily triggered by the arrests of the governor of Miyagi Prefecture and the mayor of Sendai (the capital of Miyagi Prefecture) in 1992 on corruption charges. There were deep concerns about the extent of corruption and investigations soon yielded major results, exposing long standing practices of false disbursements, padded accounts and secret funds throughout the Miyagi prefectural government offices (Sendai Shimin Ombuzuman 1999).

There are several notable points about the opportunities that have opened for social movements by the creation of the citizens' ombudsmen offices. First, through formal information disclosure processes, petitions, audits and lawsuits, evidence was forced from the government administration and, by pursuing contradictions and inconsistencies in the evidence, improprieties were uncovered and made public. Thus, while Japanese citizens' movements have long lamented the inaccessibility of government information, new information disclosure procedures and expert analyses have proven to be powerful weapons against a secretive and corrupt political system and bureaucracy. This new type of citizens' movement typically did not directly protest or petition for change (although sometimes they did both), or simply express its opposition to the government, but employed expert research and analysis methods to present incontrovertible evidence of the authorities' wrongdoings.

Second, national networking was regarded as important from the outset, and the Japanese Citizen's Ombudsman Association (JCOA) was created in 1994. Employing the successful methods employed in Miyagi prefecture, a national campaign for separate but simultaneous petitions in every prefecture and major city for the disclosure of bureaucrats' and politicians' meal allowances was submitted on 25 April 1995. The resulting disclosures revealed that approximately 80% of the meal allowances paid to the bureaucrats of the secretariats, finance departments and Tokyo offices of the 40 prefectures and major cities were for entertainment expenses, and most of that was spent entertaining bureaucrats from the national government. It is worth noting that the JCOA was formed by the initiative of groups of lawyers who had completed their legal apprenticeships together, and rapidly developed into a nationwide professional network.

Third, based on this network's internal data collection and analysis, a scale of assessment was devised, primarily judging accounting transparency. From 1997 a national ranking of local governments, all 47 prefectures and 13 major cities on information disclosure has been compiled and made public annually.[4] This can be seen as a concrete indicator of the citizens' sector's role as a critical public monitor of the government sector.

Fourth, through the above factors, processes, and networks, real results have been achieved in reforming the local political/ administrative systems nationally. The original epicenter of this activity, Miyagi, has come first among prefectures in the information disclosure ranking six of the last seven years (to 2003). Akita and Aomori Prefectures, which ranked at the bottom in the first two assessments, have made rapid progress up the ranking. In other words, the reforms have been extended far beyond information disclosure itself, and have permeated through the government finance and accounting systems, including improvements to the auditing system, and have seen new resources allocated for enforcement.

Fifth, the movement received extensive and positive mass media coverage and strong public support. The Sendai Citizens' Ombudsman has created a support organization called 'Tie Up Group' to encourage broader participation and to raise funds for various activities, such as the payment of fees to inspect government documents, photocopying, travel expenses, and filing lawsuits etc. The fact that over 300 members paid an annual membership fee of 10,000 yen is indicative of the high level of public interest and support.

Although late in coming, information disclosure procedures were finally introduced at the national level in April 2001. This is a strong example of bottom up policy/system reform, with local regions setting a precedent and the national government eventually catching up.

Public referendums

Japan's first ordinance based public referendum was held in Maki Machi, Niigata Prefecture on 4 August 1996, asking voters whether they wanted a proposed nuclear power plant or not. With a voter turnout of 88.3%, and 60.9% of them voting against the facility, this referendum set a powerful precedent for regional disputes over nuclear facilities, American military bases and industrial waste dumps throughout Japan. The Maki Machi movement inspired groups all over the country to demand public referendums, and 12 have since been held (until the end of December 2003; see chapter 9).

From 1979 until the end of September 2000, as an annually increasing trend, there have been 124 direct petitions or proposals from a governor or assembly member for a public referendum (Imai 2000). Hence, although in most cases governing assemblies are controlled by conservatives afraid of 'participatory democracy' and citizen's activism, and tend to reject petitions for public referendums, demanding a public referendum has clearly become a standard tactic for social movements since 1996.

The socio-political background and the effectiveness of this tactic were discussed in chapter 9. The cases examined there indicate that in order to get an ordinance passed through the assembly to hold a public referendum, people who support the referendum (or the movement behind it) must be elected to the regional assemblies. In Maki Machi, until April 1995 there were only two members of the town assembly who clearly opposed nuclear power. When the ordinance was passed in 1996, more than half of the seats in the assembly were held by members who were opposed to the construction of the proposed nuclear power plant. It is likely that social movements will continue to try to win seats in regional assemblies.

Policy-orientation and globalization

Collaboration and exemplary action

While confrontational movements that denounce and criticize governments and industry are stalling, there is a growing tendency for movements to develop or adopt a strong policy-orientation and work in collaboration with the government and business sectors. Citizens' movements and NPOs, as we have seen, can function as independent observers and 'countervailing power' (Galbraith 1952), critically evaluating policy proposals and determination processes. When governmental power is highly centralized in a highly technocratic system like Japan's, with economically oriented technocrats determining social policy and defining the 'public interest' behind closed doors and an enduring single party dominated political system, citizens' movements and NPOs, with their fresh ideas and flexible approaches can provide 'exemplary actions' and open the way to break existing barriers, as well as monitoring and counterbalancing the politicians and bureaucrats, especially at local level.

Collaboration involves inter-territorial, equal and specified cooperative operations performed on an individual project bases (see chapter 10). Partnership is a similar concept but refers to an unspecified, continuous relationship with implications of shared destiny. In contrast, the main characteristic of a collaboration is that it is an unprecedented type of cooperative relationship that transcends institutional barriers. Its second characteristic is that the merits and demerits of continuing the relationship are judged on a project-to-project basis.

Even in cases of confrontational issues like nuclear power and energy policy, many collaborative experiments have arisen. For example, the Tokyo Electricity Corporation and the citizens' group Renewable Energy Promoting Peoples' Forum (REPP) have collaborated since 1997 to popularize solar energy.

It is generally not possible for citizens' movements to create or enact new policies on their own. But it is possible, through open exchange between the 'specialist knowledge' of the technocrats and academics, and the 'life knowledge' or 'civil knowledge' of the citizens' sector to ensure that government policy is based upon experiential awareness of the issues.

As described in chapter 10, to strengthen the policy and business orientations of movements, it is crucial for the Japanese citizens' sector to construct a dynamic and vibrant civil society.

The information age, internationalization and the citizens' sector

Internationally, NGO/NPOs have begun to work with governments and have been key actors at large international conferences since the Earth Summit in Rio de Janeiro in 1992. For example, of the 9,850 people who formally attended the 1997 Kyoto Conference on the Prevention of Global Warming (COP3), excluding media organizations, 3,865 were from 278 observer groups including environmental and economic NGOs, both foreign and domestic.

Until the late 1980s, European environmental movements typically maintained their distance from the political authorities. However, major transformations have occurred in Europe since the late 1980s, coinciding with the collapse of the Cold War structure. With decentralization of government power and, for instance, deregulation of the electricity sector, NPO initiated environmental experiments have been conducted in numerous regional cities, with successful measures then extending across the nation and beyond. EU policy concerning global warming and renewable energy is often generated through similar bottom up processes (Iida 2000). Fortified by these successes, international NGO/NPOs with strong footings in Europe, like the World Wildlife Fund, Greenpeace and Friends of the Earth, have achieved a powerful voice in global environmental politics in recent years (Yamamura ed. 1998; Matsumoto 2001).

In the information age it is now possible to access vast arrays of information on the governments, international organizations, enterprises, NGO/NPOs, universities and research facilities of every country via the internet and personal computers—at home, in the office or in between. Until the 1990s, language barriers and spatial distances largely determined that foreign information was primarily communicated to the general public through the mediation of governments, the mass media and academics. The now routine exchanges of information between countries have opened unfathomable avenues for individuals and organizations to participate in a new global citizens' sector. Local movement strategies and goals are now continually revised and refined through these international relationships and collaborations. Simply from interaction and cooperation with European and American NGOs, with their close proximity to government and strong research and policy-orientations, perhaps Japanese NGOs may further develop their own policy influencing capacities.

Conclusion: Towards a vibrant civil society

A bottleneck for environmental movements

After more than a decade of slow economic growth, Japan retains its position as a leading economic superpower with the world's second highest GDP. Yet, in comparison to the environmental movements of other advanced industrial societies, Japanese environmental movements have had relatively little influence on either domestic policy or the international environmental movements.

Nevertheless, Japanese residents' movements, citizens' movements and environmental NGO/NPOs have come to function as non-institutional monitors of the government and industrial sectors of society, and as a countervailing force (Galbraith 1952), critically observing and challenging government policies and policy-making processes. The government's institutional structures for checks and balances have proven inadequate in the context of a highly centralized technocratic policy-making and administrative system 'overseen' by a single-party dominated political system. The supposedly deliberative bodies of the Diet and other councils of review have become ineffectual. In this context, the comparatively meager contributions of environmental movements have nevertheless had substantial effects.

In spite of the strong centralized Japanese technocracy, environmental movements can claim some major achievements. The mid 1970s saw the final stages of two decades of rapid economic growth. This period also represents a 'turning point' for social movements, as the effectiveness of the local residents' type of movements reached its limits. In simple terms, the institutions of policy formation, the political opportunity structures and the capacity of the movements to mobilize resources determined this turning point.

The closed nature of the policy formation institutions and processes limited the effectiveness of public expressions of opposition and obstructed any reforms that might enhance the power of social movements. For a long time it seemed that social movements were stuck in a vicious cycle. Regional and municipal governments were largely excluded from the planning processes for large national projects such as bullet trains, airports and nuclear facilities, thus further limiting the avenues available for concerned citizens' to participate or express their opposition to government projects, policies etc.

This one-sided closed political system encouraged state authorities to respond to public opposition through placatory strategies and, when that did not work, strong-arm tactics. Their uncompromising stance produced intense mutual distrust and obstructed the possibility of negotiated agreements between the proponents and opponents of new developments, leading to escalating costs and severe social disadvantages. In this system, active opponents to development plans were burdened with enormous time, economic, material and psychological costs. In small remote locations such costs are magnified, as oppositional advocates are typically highly visible minorities who become socially isolated or ostracized (e.g., see the case of Rokkasho discussed in chapter 9).

As discussed in chapter 7, litigation can provide a social movement with an opportunity to debate the merits of a particular development or situation in a (relatively) equal and open forum (the courtroom). But since the 1981 Supreme Court ruling in favor of the government in the Osaka Airport case, the courts have assumed a quite passive position towards public demands for them to intervene in public works, fearful of transgressing the separation of powers between the judiciary and the executive. In other words, the courts confirmed the government's prerogative to decide public policies and the 'public interest'. Thus, while it is not unusual for a European court to revoke the operating license of a commercial nuclear plant, in Japan the only similar ruling was handed down by the Kanazawa Branch of the Nagoya High Court in a lawsuit in January 2003 to stop the operation of the Monju fast breeder reactor.

Limitations of 'oppositional protest' type movements

In general, the more closed the political structures in relation to the issues concerned, the more likely it is for oppositional social movements to develop a value- or identity-oriented style of action. As discussed in chapter 9, a typical example was when the Chernobyl nuclear accident heralded an opportunity to form a new anti-nuclear movement supported mainly by urban housewives. While political opportunity structures remain closed to citizens' participation, social and environmental movements that oppose government projects are likely to remain 'oppositional protest' movements with limited capacities to present alternative policies. Japanese social and environmental movements have generally had difficulty shedding this oppositional-orientation.

Second, under these conditions, Japanese social movements typically suffer from the following internal weaknesses in their resource bases. 1. There are few national organizations and the ones that do exist are typically small with little power. 2. Specialist staff is scarce and there are few resources for internal training. 3. It is therefore difficult to cultivate an atmosphere conducive to designing creative alternatives to proposed policies. 4. Local movements are totally absorbed in local problems, impeding substantial national co-operation. 5. Horizontal communication between movements is not easy to develop, so passing a repertory of experiences, knowledge, tactics and strategies to geographically distant groups is difficult. Thus, in comparison to various European and North American countries where environmental NGO/NPOs have had a substantial influence on environmental policy, Japanese environmental movements have been less effective, largely because of their limited resource bases.

In contrast, in South Korea and Taiwan, NGOs have recently been able to dramatically influence government policy, especially in environmental protection and social welfare. Both countries have large nationwide central coordinating bodies, namely the Korean Federation for Environmental Movement (KFEM)[1] and the Taiwan Environmental Protection Union (TEPU).[2] Strictly speaking, there is no counterpart to these organizations in Japan. The activities of major international environmental organizations in Japan are more specifically focused on issues such as nuclear issues, global warming or bird protection. The membership of these organizations is relatively small, usually less than ten thousand, and their influence on national government environmental policies is quite limited.

In both Korea and Taiwan there were substantial changes to the NGO sectors immediately after the rapid democratization of the late 1980s, especially after opposition parties affiliated with NGOs and protest movements won the presidential elections in 1998 and 2000 respectively.[3] Pro-democracy activists and other citizens have joined NGOs to provide monitoring and support for the new ruling party and government.

In the Japanese political context, closed political structures have restricted the capacity of environmental movements to mobilize resources. At the same time, the typical self-framing of 'oppositional protest' type movements limited the range of goals and strategies that they could embrace. The structures of political opportunity also create a vicious circle that has been difficult to break out of. Political movements in Japan that have had substantial difficulty progressing within this structure include the 'reformist' Social Democratic Party of Japan, the Japanese Communist Party, labor movements, the Democratic Party of Japan etc. In contrast, in both Korea and Taiwan, the victories of pro-democracy parties in the presidential elections provided opportunities to break the political opportunity structure wide open. These two cases thus provide strong support to the previously stated hypotheses about the openness and closedness of political opportunity structures.

Political opportunity structure, resource mobilization and cultural framing

In recent years social movement theory has produced what I call 'the triad of social movement analyses'. From this perspective social movements are activated or strengthened by opening the structures of political opportunity, enhancing the ability for mobilizing resources and cultural framing. As discussed in chapter 5 (see Figure 5.1), this triad refers to a synthesis of the theories of collective behavior, resource mobilization and new social movements and addresses the three basic elements of social movements: discontent, collective action and orientation for change.

Chapter 9 presented a detailed analysis from this perspective, comparing the complete success of the movement in Maki Machi—which achieved the first Japanese public referendum on the question of constructing a nuclear power plant and finally succeeded in forcing the abandonment of the project—with the stalled opposition movement against nuclear fuel processing facilities in Rokkasho, Aomori Prefecture. Whereas the Rokkasho movement had great difficulty in developing broad community support, the Maki Machi movement was quite successful in this regard. The structural conditions that contributed to the victory in Maki Machi included: redrawing the electoral boundaries, resulting in a significant shift in the regional power balance (including the relocation of the long-standing LDP Diet member to another electorate); a strong capacity to mobilize resources, primarily through independent local professionals and small business leaders; and the widespread support attracted by the movement's new frame of self-determination through the referendum.

The importance of framing can be seen in the case of the 'the woods, the darling of the sea' movement initiated by oyster farmers in the town of Karakuwa Machi, Miyagi Prefecture to reforest the upper reaches of the Ōkawa River. This inclusive frame resonated throughout Japan and is now frequently mentioned in elementary and junior high school textbooks. The success of this movement provided new social opportunities for stopping the proposed Niitsuki Dam in the middle reaches of the river. Obitani's analysis attributes the success of this movement to the mobilization of external resources, which were substantially enhanced by the attractive framing of the catchphrase 'the woods, the darling of the sea' (Obitani 2000).

The institutionalization of environmental movements

In the USA, where the political opportunity structures are open, particularly in states like California and Massachusetts with well-known universities, the environmental NPO/NGOs' resource mobilization capabilities are quite strong, comprising extremely large institutional and financial support and human networking. They have therefore been able to exert a strong influence on the policy-making processes of state and federal governments. Large NGO/NPOs have offices in state capitals, from which they gather information and lobby the government. This trend has been most apparent since the success of the civil rights movement (the early to mid 1960s), but both resource mobilization capacity and influence have been progressively increasing since the 1980s. There are now more than ten major NGO/NPOs in the USA with memberships numbering in the hundreds of thousands and annual budgets of tens of millions of dollars (Mitchell et al. 1992).

As McCarthy and Zald (1977) and Perrow (1979) have observed, specialist staff in these movements, such as lawyers and researchers, play essential roles. Researchers, lawyers, journalists, consultants and other independent professionals perform an independent monitoring function in societies that are dominated by major corporations. Furthermore, when these individuals make career moves into the government or private sectors they bring their environmental awareness into their everyday contacts with new co-workers, and maintain connections with their previous networks, thus deepening and strengthening the informal interactions between these sectors. Within the movements, professional staff provide channels for communicating the opinions of the general public, translating them into the techno-legal discourse of the government bureaucracy. Similarly, they facilitate communication in the other direction, explaining technical language or the legal system in plain language that the general public can readily understand. The more influence that NGO/NPOs have on the policy-making processes, the more easily they can garner support from the public, raising funds through corporate and individual donations. With funding they can employ full-time specialist staff, thereby strengthening their capacity to contribute to policy-making. These are the opportunities that open for social movements when institutionalized—the capacity to realize the principles of the 'resource mobilization theory'.

At the same time, the institutionalization of environmental movements and their growing political power entails an increasing focus on the organization's self-interest as a pressure group and its cooptation into the system it set out to reform. Here it is important to remember that these new institutions were called into being to redress the shortcomings of an overly institutionalized and unresponsive political-economic-social system (Terada 1998a; 1998b).

Expectations of the citizens' sector

Since the 1992 Earth Summit, in many countries national governments and NGO/NPOs have maintained a dialogue through international institutions, which are typically the central players at large international conferences. Matsumoto Yasuko (2001) vividly describes the strategies used by international environmental NGOs in their negotiations between nations, based on her own experiences as the head of Greenpeace Japan's global warming and CFCs campaigns.

Against a background of 'government failure', 'market failure' and 'communal failure' (see chapter 11), international expectations for the non-government, non-commercial, 'third sector' or 'citizens' sector' of the NGO/NPOs are rising. The citizens' sector is inclined towards the creation of a 'sustainable society', forming a new, open, public sphere and reforming various practices related to public issues. As discussed in chapter 11, the opportunities for the citizens' sector in Japan to realize their objectives are constrained by the peculiar conception of the 'public', the weakness of public philosophy, the institutions and processes of the hypertrophied state, and the interests and machinations of the mass media etc.

Effects of NPO incorporation

Although international NGOs were already prominent before the 1992
Earth Summit, it was only around this time that a push began in Japan
to give citizens' sector organizations corporate status. Following the
Hanshin earthquake there was a sudden surge in public awareness
of the social mission and necessity of civil action. The NPO law was
soon passed, giving corporate status to twelve (recently expanded
to fifteen) different categories of citizens' groups. Environmental
conservation was one of the original twelve categories, and 28.1%
of all incorporated NPOs cited 'actions seeking environmental
conservation' in their articles of association, making it the seventh
largest category of activity, as indicated in Table 14.1.

Incorporation is of course both a mode and an element of
institutionalization, which as we have seen has both risks and
benefits for a citizens' movement. At the least, incorporation entails
fundamental transformations for a citizens' group. As mentioned,
Japanese environmental movements until the 1990s were typically
oppositional protest movements, critical of industrial development
in general and development-oriented government administrations
in particular. They were typically reactive movements, responding
to individual problems after they arose. Often, when the immediate
dispute died down, the movements stagnated or dissolved altogether.

Table 14.1 Categories of activities of incorporated NPOs

Categories of Activities	Number of Incorporated NPOs	Ratio
Health and medical care and welfare	4956	59.6%
Social education	3740	45.0%
Intermediary of organizational management	3229	38.8%
Urban policy	3087	37.1%
Childcare and nursing	3054	36.7%
Culture, arts and sports	2441	29.4%
Environmental conservation	2337	28.1%
International cooperation	2019	24.3%
Human rights, peace and stability	1293	15.6%
Gender relation	794	9.5%
Community safety	692	8.3%
Disaster relief service	636	7.6%

Note: Tables compiled from Cabinet Office data current at the end of September
2002. Each organization may appear in multiple activity categories. The ratio column
indicates the percentage of the total of 8315 corporations engaged in a particular
activity at the time.

In other words, they were one-off, single-issue movements. To be candid, many of these movements were expressions of local self-interest, or nimbyism, compelled by a sense of impending peril into defending the material conditions of their lifestyles or their means of production (for example, farmers and fishermen), desperately attempting to protect themselves and their regions from immediate threat. To that extent, they rose quickly and then faded away after a short period.

An incorporated citizens' movement, however, can employ permanent, full-time staff and establish a stable and well-resourced office from which to proactively and routinely address a range of environmental problems. Thus incorporation may extend the life of a movement, which may then shift its focus from an after-the-fact reactive movement to a proactive prevention-oriented movement. Perhaps we may anticipate the growth of more policy oriented environmental movements—born at the local level, then growing nationally, expanding to the East Asian region, and finally having global effects.

Grassroots environmental NPOs and international NPOs

In recent decades there has been rapid growth in the number of local citizens' groups involved in recycling campaigns, a trend that can be expected to continue. Corporate status is both necessary and highly advantageous for such organizations, enabling them to secure premises and equipment, and to form contracts with merchants and traders. Corporate status is also advantageous for consumer groups, those looking to promote reforestation, the purification of waterways or to popularize renewable energy resources. With support from junior chambers, consumers' cooperatives, agricultural cooperatives, local media and others, citizens' groups that value cooperation and collaboration with local governments will seek corporate status, because incorporation as an NPO provides leverage and legitimacy in networking and developing collaborative relationships. NPOs that have a strong grassroots type, non-professional character, greatly enhance their influence and thereby the potential for expanding their activities when they establish a strong base within a local community.

Another type of NPO is the highly professionalized environmental NPO located in metropolitan areas. While in Japan these principally operate on a national level, they also seek to expand their activities to an international level, providing a center for activity in the East

Asian region, for example. At the national level environmental NPOs perform the function of alternative think tanks and consultancies that counter government and corporate think tanks. Japanese branches of international NGOs like the WWF and Greenpeace appear set to strengthen this side of their operations. Kiko Network (formerly Kiko Forum), CASA (Citizens' Alliance for Saving the Atmosphere and the Earth) and Citizens' Nuclear Information Center are some of the environmental NPOs originating in Japan that have achieved international results.

How the public sphere can be opened

It is worth considering whether the current growth in environmental movements is merely a passing fad created by the NPO boom, or has the potential to create enduring and effective institutions. The citizens' movements that sprang-up in nimbyish reactions to local threats tend to grow rapidly because of the urgent need for action—if they did not protect their own 'backyards', their lives and livelihoods might be irreversibly changed by development and pollution. Will the most recent NPO boom in social movements and citizens' groups also fade away?

It is clear from the extensive participation and financial con- tributions of the general public that NGO/NPOs around the world are being called upon to at least partly fill the vacuum in social services and the political checks and balances produced by the triple failures of the government, market and community—to replace the government with its inadequate and sometimes over-extended reach, the too conservative courts with their restrictive precedents, the deteriorating labor movements and political parties co-opted by their pursuit of particular interests, and the various facades of civic and autonomous councils. The central question ahead is how best to open the systems for addressing public issues to the silent majority of 'ordinary people'—that is, how to create a new public sphere.

The first answer to this question is to focus on devising and performing/proposing exemplary actions, thinking laterally to find existing and create new opportunities to break through the various blind alleys in which environmental problems are frequently stuck. While collaborative relations with government and enterprise are strained, by strengthening the capacity to perform a social observer role, investigate environmental problems, and propose and critique policies, innovative experiments can yield ground breaking results in regional cities, as the European environmental NGOs have clearly demonstrated. Successful collaborations carried out between NGO/ NPOs and local/regional governments are often adopted across the nation, and may then be adopted at local, regional and international levels. As explained in chapter 10, after the Hokkaidö Green Fund's resoundingly successful share-offer to raise funds for a green power scheme, similar undertakings are rapidly spreading across the nation. This seems to suggest that projects with strong social benefits that help to realize peoples' dreams of participating and contributing to the creation of the public good are enormously attractive.

Exemplary actions arising through dialogical civil interaction is the primary mission and *raison d'être* of reform-oriented social and environmental movements.

Regional possibilities

The second answer is a keyword of this book: collaboration. Collaboration is the performance of tasks through relationships that are: equal, inter-territorial, single-project oriented and open. Among the necessary conditions for collaboration are the identification and definition of mutual objectives and shared benefits. Direct interactions between participants and trust-relationships are fundamental to collaboration.

The third answer is regionally initiated reform. At the local and regional levels of government in cities like Sendai, with populations of up to around one million people, the social distance between citizens and government officers is relatively small. Some government officers, acting in their capacities as private citizens on their holidays or after hours, are increasingly supporting residents, citizens' movements and NPO activities. On a tide of sweeping regional decentralization, the substantial roles that local govern-ment employees can fulfill are gradually expanding. As Tanaka observes, when viewed nationally, even in the energy field where regional authority is very small, various trial projects have been initiated in Japan to advance the use of natural energy resources, each appropriate to the particular regional circumstances (Tanaka 2002).

As mentioned in chapter 12, Japan does not have small uni-versity cities comparable to Davis, California. But in centrally located regional cities with populations in excess of two hundred thousand people, there is little social distance between independent professionals—including academics and school teachers, graduate students, lawyers, media professionals, medical practitioners etc—and local government employees, corporate employees, business leaders and so on. It is therefore relatively easy to form collaborative relationships. Evidence of the steady growth of collaboratively formulated responses to public issues is now clearly apparent to the ordinary observer 'in the streets', even in small Japanese cities. This is the activity that is creating and defining a new public sphere. These Japanese cities have sufficient resources to be proving grounds for public and environmental policy.

Even in local towns and villages with populations of around ten thousand people, where it was once thought that citizens' requests for dialogue or resources would be rejected, when local government employees and leaders, local volunteer groups and others work together with independent professionals—from near or distant cities if necessary—the region's small scale may facilitate successful social experiments, whose very success then has a multiplier effect in stimulating further experimentation. For example, even in Yamagata Prefecture, considered to be a very conservative area, Tachikawa Machi successfully launched a regional wind power generator (Hasegawa 1998e). Takahata Machi is nationally recognized as an advanced organic farming region due to the success of its alternative distribution system based on mutual understanding (derived through direct communication) between producers and consumers (Matsumura and Aoki eds, 1991). In Kanayama Machi, an area recognized for its cedar tree-farms, rows of beautiful wooden houses constructed of the local cedar have been restored. The city of Nagai adopted the slogan 'Connecting farm and kitchen' to increase participation in its household waste collection and composting project, oriented towards encouraging organic farming (Aoki 2001). These are but a few of the cases of regionally developed and oriented environmental policies that have achieved national recognition and acclaim.

In Europe, the United States and Japan, regional cities, towns and villages have become the principal stages for the exemplary actions of environmental movements and the primary loci of the new public sphere. Herein lies the basis and the *raison d'être*, of environmental sociology, too.

Notes

Chapter 1

1. Based on the calculations of Funabashi Harutoshi.
2. For representative examples, see Iijima (1970b), Matsubara and Nitagai (eds) (1976) and Motojima and Shōji (eds) (1980).
3. Based on information provided by the former representative of the Section on Environment and Technology, Professor L. Lutzenhiser (Washington State University) and the secretariat of the American Sociological Association. The American Sociological Association has a membership of about 13,000 and, as of 2002, the three sections with the largest membership were Sex and Gender (1064 members), Medical Sociology (1003) and Organizations, Work and Labor (985). Environment and Technology has a membership of 409 people, making it the 21st largest of the 43 sections.
4. RC 24 was called Social Ecology at the Madrid Conference in 1990, but combined with the working group on Environment and Society during the Bielefeld Conference in 1994 and changed its name.
5. Based on data supplied by Professor Mol (Wageningen University), president of the Environment and Society Research Committee (perscom).
6. For the main joint research projects, see Iijima (2002) and Iijima Nobuko Sensei Kinen Kankō Iinkai (eds) (2002). For a representative work, see Iijima (ed.) (2001).
7. For the concept of 'environmental coexistence' in environmental sociology, see Torigoe (2001a). The life-environmentalism proposed by Torigoe and his colleagues introduces this as a counter-concept to 'environmental problem' or 'environmental degradation'. For example, they found that the traditional water management system or land use system in the area surrounding Lake Biwa was a 'coexistent relationship', embedded in the environment, with practices stressing natural cycles and conservation mechanisms (see Torigoe and Kada eds, 1984 and chapter 2).
8. *Environmental sociology in Japan* (*Kōza kankyō shakaigaku*) consists of: volume 1, *Perspectives of environmental sociology* (*Kankyō shakaigaku no shiten*), volume 2, *Environmental perpetration, suffering and the process of solving environmental problems* (*Kagai higai to kaiketsu katei*), volume 3, *The natural environment and environmental culture* (*Shizen kankyō to kankyō bunka*), volume 4, *Dynamism of environmental movements and policy* (*Kankyō undo to seisaku no dainamizumu*), and volume 5, *Asian and global environmental problems: Sociological studies of local origins* (*Ajia to sekai—chiiki shakai kara no shiten*).
9. See Kaku Kensuke (1996).
10. Based on comparative research of Germany and Japan, G. Foljanty-Jost argues that Japanese environmental policies were internationally advanced in the 1970s, but Japan lost its pioneering status from the 1980s. She argues that this was because environmental NGOs were relatively weak and were not given access to the environmental policy-making processes (Foljanty-

Jost 2000). More recently there has been extensive comparative sociological research into environmental policies (Münch et al. 2001) and comparative research from a political-sociological perspective on trust in politics (Pharr and Putman (eds) 2000).

11. At the Montreal Conference in 1998, four papers were given by Japanese researchers in the Environmental and Society Research Committee: Aoyagi Midori (joint research with Kuribayashi Atsuko), Iijima Nobuko, Mitsuda Hisayoshi and the author. At the Brisbane Conference in 2002, there were four: Aoyagi Midori (again joint research with Kuribayashi Atsuko) and Terada Ryōichi giving one each, and the author two.

12. The Japanese Association for Environmental Sociology established an International Exchange Committee from 2002 to tackle this problem.

13. It is possible to do a keyword search of the *Shakaigaku bunken jōhō dëtä bësu* (Bibliography of Japanese Sociology Database) on the Toyama University homepage (http://jinbun 1.hmt.toyama-u.ac.jp/Socio/jss/index.html, last accessed 31 January, 2003), which has 64,000 pieces of data on sociological research published in Japan from 1945 to 2001 (title and subtitle included). The author did a search for 'environment and municipalities', and had only ten hits. Even 'environment and city' had only 88 hits, and this included titles such as 'Elderly residents in the city and the living environment'. Judging from the titles and authors, there are only 40 or so entries on the natural environment, historical environment, and environmental protection. 'City' alone produces 3536 hits and 'environment' produces 940 hits. This indicates how little intersection there is in Japan between research in environmental and urban sociology. There were 401 hits for 'city and life', and 201 for 'city and residents'. Urban sociology has a 'human-centered paradigm', which indicates the assumption that the city is relatively complete and autonomous. See Hasegawa (2003). (Note that there are some cases where the documentary data is duplicated in the database. In cases where there were less than 100 hits, the author removed all cases where the data was duplicated, but where there were more than 100 hits, there may be some duplication.)

14. In his late years, Parsons (1978) attempted to incorporate the natural environment into his social systems theory, but did not make any concrete progress. See Tominaga (2002).

15. Yonemoto (1994: 256) ends his book with the words 'already somewhere on this planet, it is possible that someone has begun to write a book that will play the role of Marx's *Capital* in environmental issues'.

16. The word 'environment' comes from the French word *viron*, meaning a circle, a round and surroundings. Abstractly if you can define the system, all entities surrounding or outside of the system belong to its environment.

17. 'Semi-environment' refers to a nature that has been irrevocably affected by humankind. See Torigoe (2001b), Miyauchi (2001), and Tokuno (2001).

18. For the main research trends in environmental sociology, see Iijima (1998b) and Horikawa Saburō (1999).

Chapter 2

1. The structure and features of different types of pollution and environmental problems are analyzed from the sociological perspective in chapter 3.
2. This case is examined in detail in chapter 9.
3. See Funabashi, et al. (eds) (1998).

4. The 'New Ecological Paradigm' was called the 'New Environmentalism Paradigm' at the time Catton and Dunlap's first work was published (1978). Later, however, they revised the term to stress the ecological basis of human society (Dunlap and Catton 1979). For Dunlap's response to his critics and for his general views on the significance of the 'New Ecological Paradigm', see Dunlap (2002).
5. See Miyamoto (1999) and Nagai et al. (eds), (2002).

Chapter 3

1. Sendädo (ed.) (1991). Katö and others who played a central role in editing this work cooperated with Sendai City to edit another work, Sendai Miyagi NPO Sentä ed. (1999), which introduces 504 organizations that replied to a questionnaire sent in November 1998 to 1,293 citizens' action groups within Sendai. Of these, the activities of 31 organizations focused on the 'environment'.
2. Hidaka (1973) examines both the contents of the concept of 'citizen' in the historical and political contexts of Japanese social movements and discusses various critical interpretations of this concept.
3. The 'four major' cases of industrial pollution-caused community illnesses are the most serious and symbolic incidents in the Japanese history of environmental hazards, including Minamata mercury poisoning, Niigata mercury poisoning, Ouch-ouch (Itai-itai) cadmium poisoning and Yokkaichi asthma. Each initially appeared in the 1950s and had long histories of public denials of the existence of serious illness from industrial pollution by the companies, the authorities and even the communities where the victims were living. See Ui (ed.) (1992) and Broadbent (1998: 103).
4. See Hardin (1968) and Umino (1991). The social dilemma is defined as a dilemma or conflict between collectively and individually rational action, where the action required for achieving the collectively best outcome or goal is in conflict with the action required for achieving the individually best outcome. The Prisoner's Dilemma, 'tragedy of commons' and 'free rider problem' discussed by Olson (see chapter 4) are typical examples.
5. Hasegawa (1998b) documents my own participation in and observations of the Kyoto Conference. Issue four of the *Journal of environmental sociology* (*Kankyö shakaigaku kenkyü*) contains the accounts of four female NGO leaders who participated in the Kyoto Conference. See also Asaoka (1998), Matsumoto (1998), Ayukawa (1998) and Höjö (1998).

Chapter 4

1. See Kimura (2002) for a critical analysis of the size factor in Olson's conditions.
2. This scenario is based on the author's own case studies of the construction of bullet train lines and nuclear power related facilities, see Funabashi et al. (1985) and Hasegawa (1998a).
3. For a detailed discussion of the status of and issues faced by internationally active Japanese and foreign environmental NPOs and NGOs, see Yamamura (ed.) (1998).
4. As a result of the revision in 2001 of the Special Taxation Measures Law, a 'preferential tax treatment system for recognized NPOs' was introduced. Under this system, in cases where individuals or firms make donations

to recognised NPOs, they are allowed within certain limits to deduct the donation from taxable income. However, the conditions that have to be met to have a recognized NPO as a juridical entity are very strict, such as having at least one third of total revenue coming from donations and subsidies. Only nine groups were recognised in the first year after the system was introduced (Yamaoka 2001).
5. For a detailed discussion of the financial support for American NPOs, see Okabe (2000).
6. For the concrete roles performed by NPO Support Centers and intermediary organizations in Japan, see Li (1999). For the contents and history of activities, see Sendai Miyagi NPO Sentä (2002).
7. *http://www.jnpoc.ne.jp/support/index.html*, last accessed 31 October 2002.

Chapter 5

1. For an analysis of the intersection between environmental sociology and social movement theories, see Takada 1995.
2. Many researchers have emphasized the qualitative changes in environmental issues due to the recent diversification of environmental problems and the growing debate on global environmental problems. However, I have proposed that industrial pollution, high-speed transportation pollution, everyday life pollution, and the global environmental problems be seen as a single unified set of issues, through a definition that moves beyond superficial differences and states that 'environmental issues are problems concerned with the emission of and process of dealing with downstream environmental burdens that are caused by upstream production processes and life processes' (see chapter 2).
3. The term 'solution' strongly suggests the concrete resolution of individual environmental problems. In the case of global environmental issues, the term 'solution' is frequently inappropriate; 'improvement' or 'correction' should be used instead. The term 'policy studies' refers to (1) generalities and well established fields of institutional analysis and policy research, (2) current issues such as global environmental problems, and (3) the orientation of environmental economics and environmental legal studies, which are policy-oriented rather than solution-oriented.
4. For a detailed discussion of the current status of and issues facing Japanese and other environmental NGOs and NPOs that are active on the international stage, see Yamamura (ed.) (1998) and Matsumoto (2001).
5. In social movement research, the area that is most wanting is research into the effects and social consequences of actual movements. For recent re-search on social movement and institutional reform, see M. Giugni et al. (eds) (1998).

Chapter 6

1. Hasegawa Kōichi (ed.) (2001) was the first monograph published in Japan that was an academic work in environmental sociology using the term 'policy' in the title. There are few similar examples published outside Japan.
2. One of the themes of the 2000 Commemoration Symposium of the Japan Sociological Society's Annual Conference (held at Hiroshima Kokusai Gakuin University) was 'is sociology useful?'. This was the first time the

Japan Sociological Society's Conference had held a symposium on this theme.

3. Active attempts to develop policy-oriented research are limited to the joint project undertaken by Funabashi and myself on bullet train pollution and the issues of the construction of bullet train lines (Funabashi et al. 1985, 1988; Funabashi 1990, 2000, 2001), Tanaka's work on river administration (Tanaka 1997), and my own work on nuclear policy (Hasegawa 1996c, 1999a).

4. Nishihara et al. conducted a trend-analysis of the articles published in the Japan Sociological Society's official journal *Japanese sociological review* (*Shakaigaku hyōron*) for the first fifty years after it was founded. They found that 20.4% of all articles published in the 1990s included the name of either a sociologist or a school of thought in either the title or sub-title. These were mainly the names of Western sociologists or a Western school, rather than Japanese sociologists or schools. Although this figure was down from 28.1% in the 1980s, it continued to be the largest single category of titles, and was merely a return to the levels of the 1960s and 1970s (Nishihara and Sugimoto 2000: 319).

5. Data on Greenpeace and WWF membership by nations are from Matsumoto Yasuko (2001: 187).

6. At the Conference of the Parties 6 (COP 6) talks on the prevention of global warming in The Hague, the Netherlands, in November 2000 the protesters had used sandbags to create a symbolic floodwall to warn against the damage of rising sea levels as a result of global warming. When the Prime Minister of the Netherlands gave his speech, he stated that people the world over are sincerely concerned about whether or not the conference would come to an international consensus to protect against global warming. He placed one of the sandbags on the stage, and said that it was a symbol of the worries of the world's citizens. The sandbag remained there for the remainder of the conference as a gentle reminder that the world was watching. In Japan it is hardly imaginable that the Prime Minister might refer to protesters' activities in any positive manner. The author has discussed this impressive episode in Hasegawa (2001b).

Chapter 7

1. There were virtually no Japanese academic works from the social movement perspective on the 'Lawsuits strategy' when this chapter was first published (1985). Iijima (1970a), Ui (1974) and Matsushita (1980), (the latter two were both movement leaders rather than professional sociologists) were written in a more practical style to stress ideals of the movement. Works in the sociology of law from the same period include Sax (1971), Tanase (1972), Awaji (1973, 1980), Tanaka (1979), and Toyoda (1982). Research after 1985 includes Miyazawa (1989, 1994) and Ōsawa (1989) (the latter compared Japan and the USA).

2. Refer to Nagoya Shinkansen Kōgai Soshō Bengodan (Nagoya bullet train pollution lawsuit plaintiffs' counsel) (1996), Osaka Kūkō Kōgai Soshō Bengodan (Osaka Airport pollution lawsuit lawyers' group) (ed.) (1986), Kimura and Kuboi (1978), and Kimura (1982).

3. The original version of this chapter (Hasegawa 1985b) and Funabashi et al. (1985) received a strong response from legal sociologists such as Miyazawa (1989, 1994). Hasegawa (1989) was based on a paper presented at the 1988

Japanese Association of Sociology of Law Conference. See also Rokumoto (1991: 4).

I have also critically analyzed the Iwate anti-Yasukuni Shrine lawsuit from a similar perspective (Hasegawa 1993). Nishio (2002) applies Melucci's (1989, 1996) social movement theory to judicial action for a principle of separation of religion and the State.

4. The Osaka Airport pollution lawsuits consist of five successive suits with largely the same frameworks and plaintiffs' counsels. The first suit was filed in 1969 as a test case by the 31 plaintiffs identified as suffering the most severe damage from airplane noise. Gradually the number of plaintiffs grew larger: the second suit was filed in June 1971 by 124 plaintiffs and the third was filed in November 1971 by 109 plaintiffs. These three lawsuits were immediately combined in the District Court and ruled upon in February 1974. This ruling awarded compensation but only a portion of the injunction demanded in the application. An organization of discontented plaintiffs' soon appealed and 10 months later filed a fourth suit as a sanction to this ruling by 3694 plaintiffs. After the Supreme Court overturned the High Court's decision in favor of the plaintiffs and dismissed the injunction claim altogether, 134 plaintiffs who had been previously involved in one of the first, second or third suits, filed a fifth suit to claim additional compensation. Combined, the plaintiffs of the fourth and fifth suits agreed to accept compensation of 1.3 billion yen and a commitment to continue the 9 pm curfew in accordance with the injunction applied for in March 1984.

5. The issue of the military base noise pollution is structurally quite similar to other airport noise pollution. But generally the damage is more severe in the military base cases and the litigation for plaintiffs is more difficult to win. Military aircraft fly lower and on irregular schedules that do not comply with strict regulations such as a 9 pm curfew. Disputes over the constitutionality of Japan Self-Defense Forces and the Japan-U.S. Security Treaty invariably complicate the issues.

6. In October 1981, there were 67 current actions for environmental/ anti-pollution injunctions, of which 49 were civil litigation, and 18 were administrative litigation ('Towareru "kökyösei"—Osaka Kükö Soshö no 12 nen ('Publicness' questioned—12 years of the Osaka Airport Lawsuit) No. 3, *Mainichi Shimbun*, 12 December 1981). Major airport pollution lawsuits filed between the 1970s and early 1980s include: Fukuoka Airport (filed March 1976, 368 plaintiffs), and Niigata Airport administrative lawsuit (filed March 1979, 4 plaintiffs, first trial August 1981, appeal court December 1981, both actions dismissed). Environmental lawsuits were filed against military bases at: Yokota (filed April 1976, 153 plaintiffs), Atsugi (filed September 1976, 93 plaintiffs), Kadena (filed February 1982, total plaintiffs of 1st and 2nd lawsuits: 906), and Komatsu (filed March 1983, 318 plaintiffs). Bullet train lawsuits: an administrative lawsuit by Edogawa Ward and its residents demanding that approval for the Narita bullet train construction project be overruled (filed April 1972, 22 plaintiffs, first trial December 1972, appeal court October 1973, and Supreme Court December 1973, all actions dismissed; later this project was abandoned for mainly economic reasons), an administration lawsuit by residents of Urawa City and its surrounds, demanding that the approval of the Töhoku and Jöetsu bullet trains construction project plan alterations be overruled (filed April 1980, 89 plaintiffs, action dismissed November 1983), a civil lawsuit by Kita Ward residents demanding an

injunction against the same construction project (filed September 1980, 203 plaintiffs, out-of-court settlement finalized October 1984).

7. I learned a great deal about the roles of the lawyers from interviews I conducted with Teruo Takagi, secretary of the Nagoya bullet train plaintiffs' counsel, and Kazumasa Kuboi, secretary of the Osaka Airport plaintiffs' counsel (at the time of the interviews. At the time of writing in 2002 Mr. Kuboi was the president of the Japan Federation of Bar Associations). See also Kimura and Kuboi (1978), and Kimura (1982).

8. The concept of environmental rights was first proposed by lawyers Hajime Nitö and others of the Osaka Bar Association at the annual meeting of the Japan Federation of Bar Associations in Osaka, 1970. Mr. Nitö was the first lawyer consulted by representatives of a neighborhood organization in Kawanishi City about filing a lawsuit. Many of the members of the environmental rights research group of the Osaka Bar Association were formerly members of the plaintiffs' counsel for the Osaka Airport lawsuit, including Mr. Kimura and Mr. Kuboi. Refer to the Environmental Rights Research Group of the Osaka Bar Association ed. (1973).

9. A lawyer, Akio Tomishima said at a round-table discussion 'legal theory is consolidated through trials' on the basis of his involvement in the Yokkaichi pollution trial (in Gö et al. 1972: 51).

10. Incidentally, the written evidence lists for some cases are so substantial that they can be used for researching relevant pollution issues. For example, the 'Documentary Evidence List' in *Höritsu Jihö* (Law journal) (1973:78–84) is subtitled 'A reference material list for airport pollution issues'.

11. The plaintiffs' group of the Nagoya bullet train pollution lawsuit received a provisional execution of damage compensation money in the first trial of about 530 million yen and used its interest equivalent to finance the appeal. This halved the plaintiffs' group's monthly fees from 1000 yen to 500 yen.

12. Councils for civil litigation trials were held in November 1976 and November 1977. The Supreme Court opinions on these councils were distributed as the 'Reference material for judging anti-pollution injunction trials' (confidential document) to each district court in March 1978. Of the major pollution cases before and after this distribution, there were 19 wins and 21 losses for the plaintiffs up to 1977, and only 2 wins against 15 losses between 1978 and 1981 (Toyoda 1982: 441–4).

13. For the Nagoya bullet train lawsuit, the cost born by each plaintiff up to the conclusion of the first trial was around 200,000 yen. The total cost for the entire plaintiffs' group was more than 80 million yen (from survey data).

Chapter 8

1. When the first version of this paper was published in January 1991, there were 39 nuclear power stations operating in Japan and their total output was ranked fourth in the world after the USSR.

2. While strictly speaking the term 'anti-nuclear movement' refers to both the anti-nuclear *power* movement and the anti-nuclear *weapons* movement, as will be explained shortly, this chapter is concerned only with the former. However, to avoid the repetition of the awkward phrase 'anti-nuclear power movement' I will refer to it as the 'anti-nuclear movement' throughout this chapter.

3. I previously compared new social movements theory and resource mobilization theory, focusing on their complimentarily (Hasegawa 1990). My recent development of this relationship is illustrated in Figure 5.1, 'Triangle of Social Movement Analysis', chapter 5.

4. This research project was funded by part of a grant from the Japanese Ministry of Education on the social disputes surrounding the construction of nuclear fuel processing facilities and the Onagawa Nuclear Power Station (Miyagi Prefecture). See Hasegawa 1991a for a summary of the research results.

5. For example: Kubokawa Chö in Köchi Prefecture, Hikigawa Chö and Hidaka Chö in Wakayama Prefecture, Iwaishima in Kaminoseki Chö in Yamaguchi Prefecture. In chapter 9, I describe Maki Machi's case in Niigata Prefecture.

6. Two notable exceptions are Reiko Watanuki and Taeko Miwa who were very active both domestically and internationally.

7. An *Asahi Shimbun* public opinion survey of residents in Aomori Prefecture in November 1990 asked if the facilities proposed for Rokkasho should be built. The results were markedly different to the response to the same question in September 1984, before the project was approved by the prefectural government.

	September 1984	November 1990
yes	31%	16%
no	35%	62%
other	34%	22%

8. Waseda Daigaku Bungakubu Shakaigaku Kenkyūshitsu (Waseda University, Sociology, School of Literature, Arts and Science) 1988.

9. McAdam (1998) reports on a prospective study of the influence that participation in the 'Freedom Summer' campaigns (1964) had on political socialization. In Japan, Kurita (1989) provides a quantitative analysis of the later effects of adolescent political socialization experiences, surveying readers of citizens' movement newsletters.

10. According to the 'Education Ministry Statistics Digest' (Monbu Tökei Yöran) (1990 edition), the proportion of women entering tertiary education exceeded 20% for the first time around 1965, when the first baby boomers reached the tertiary entrance age. In 1965, the number of female junior college students exceeded 100,000 and female 4-year university students 150,000 for the first time.

11. From an interview with a member of the 'Mothers group to protect children from radiation' (23 March 1990, Hirosaki City, Aomori Prefecture).

12. As per the interview in note 11, and a roundtable discussion held by the All Hokkaidö Opinion Advertising Coalition (1988: 61–75).

13. According to 'Thinking about the Citizens' Movements' (Shimin Undö o Kangaeru), Iken Kökoku Zendö Renrakukai (All Hokkaidö Opinion Advertising Coalition) 1988: 90.

14. In the dialect of the Tsugaru district, 'maine' means 'no', or 'don't like'. 'Odazunayo' comes from the dialect of South Sanriku district and means 'don't kid me'. Both can mean 'stop that', and are commonly used for scolding children.

15. According to the interview cited in note 11.

Chapter 9

1. Informal/unofficial referendums had previously been held on occasion by protestors. An official referendum requires a local ordinance to be passed before it can be held.

2. I have elsewhere discussed the processes through which a nuclear reactor was shut down immediately and decommissioned in Sacramento California through a citizens' initiated referendum and the social impact of the plant's closure on both the local communities and the electric utility. The effects on the utility led to its wholesale reorientation to become a model energy company, stressing energy efficiency and the use of solar energy (Hasegawa 1996c). Takubo (1996) provides case studies of the processes involved in drafting and enacting the California Nuclear Safety Act.

   Numerous sociologists have produced case studies on Maki Machi's referendum, including Takubo (1997), Yamamuro (1998), Sung (1998), and Watanabe (1999). Itö (ed.) (2000) provides the most systematic survey of the local residents.

   Sung (1998) suggests that the combination of growing risk recognition and demands for self-determination in relation to risk are the key factors for explaining the increasing popularity of regional referendums. His analysis of these factors does not, however, fully explain why referendums have suddenly become central strategies for social movements.

   Takubo (1998) was the first of the studies cited to employ the analytical framework used in this chapter—that is, the structure of political opportunities, mobilization structures, and cultural framing. Takubo primarily analyzes the town's internal political dynamics, focusing particularly on the 'Group to bring about a local referendum' as the main protagonist. The narrowness of this perspective, however, leaves Takubo's analysis of cultural framing a bit weak. In this chapter I aim to overcome this problem by outlining significant developments in the external environment and the strategic importance of structures and framing.

3. A proposed nuclear reactor at Öma Machi, Aomori Prefecture, later joined Maki No. 1 as an officially approved yet stalled reactor project. The Öma reactor was adopted into the national energy policy in August 1999, but its safety assessment has been suspended since October 2001, awaiting the acquisition of the necessary land.

4. A portion of this land had previously been the center of a long dispute between the town hall and a nearby Buddhist temple. The District Court ruled in February 1986 that the 1,176m² former graveyard belonged to the town. The High Court in Tokyo upheld the decision in October 1987, as did the Supreme Court in October 1989.

5. The construction of the Kashiwazaki Kariwa nuclear power plant, the world's largest nuclear power station, with seven reactors and a total power output of 8.21GW was only possible because of the political stability of Niigata Prefecture. Construction of the No. 1 reactor began in 1978 and commercial operations began in 1985.

6. Based on interviews with the participants of the independent (unofficial) referendum, Yamamuro (1998) refers to shigarami, or a 'compelling force'— the local social code of behavior that suppresses individual thoughts and attitudes. He concludes that the 'Group to bring about a local referendum's success lay in its close attention to shigarami'. The other side of shigarami

is kakotsuke, an acceptable social pretext for social dissent. The group and the unofficial referendums, in other words, provided an acceptable social pretext and means for citizens to anonymously express their secret intention (honne) to oppose the proposed nuclear plant. This analysis of the local residents provides important insights—especially of the actions of farmers, merchants and crafts people—but as I have argued in this chapter, we must not overlook the structural changes in political opportunities. Long-standing social control mechanisms had been severely weakened by independent events in the national and regional power structures, events that led to the relaxing of local shigarami.

7. When first proposed in 1969, the Mutsu Ogawara development project aimed to construct the largest petrochemical industrial complex yet in Japan. However, after the oil crisis of 1973, the petrochemical industry turned its attention to surviving with reduced levels of production/supply and this plan was effectively abandoned. Nuclear fuel processing facilities were invited to utilize the vacant wasteland by the Aomori Prefectural government in 1983.

Chapter 10

1. The cost of wind-generated electricity has dropped sharply in recent years due to the increasing size of the plant. The latest wind farm built in the US can generate power for less than $0.03/kW, cheaper than natural gas or coal fired thermal power (American Wind Energy Association, *Global Wind Energy Market Report, http://www.awea.org/faq/global2000.html*, 20 May, 2001).

2. Iida defines the 'green electric power scheme' as fulfilling the following three conditions: it aims to increase the use of renewable energy; it is run by the utilities or otherwise connected to power bills; and consumers can participate both voluntarily and easily (1999: 31). But, as discussed shortly, green power certificate systems and levies are also often called green power schemes, so the second and third criterion may be too restrictive. The first criteria alone, on the other hand, could include government subsidy schemes and may be too loose. Hence my definition in the text.

3. Consumer goods, shōhizai in Japanese, are commonly described by three Chinese characters, the first two meaning 'consumption' and the last, 'goods'. The Seikatsu Club, however, specifically opts to use a final character meaning 'material' instead of 'goods', stressing that the use value invested or produced by the consumer him- or herself takes precedence, rather than the exchange value of goods supplied by mainstream industry.

4. Sakae Sugiyama, the Hokkaidō Green Fund's president and Tōru Suzuki, its chief of administration, deserve most of the credit for working out the details with the utility and financial institutions and putting it all together.

5. See Tōru Suzuki (2002) and the web pages of the Hokkaidō Green Fund (*http://www.h-greenfund.jp*, 31 October 2003).

6. KEPCO has reduced its monthly charge to 100 yen since April 2000 due to the poor response, but the number of participants remains at the same low level.

7. After a split in the ruling party, LDP (Liberal Democratic Party) in 1993, many new parties emerged, faded away, merged, or were reabsorbed into the LDP. For example, the New Frontier party (Shinshin Tō) was founded in 1994 by Ichirō Ozawa, a former LDP leader, as an influential new party

including Kömeitö. It became a major opposition party in a conservative-middle camp coalition. But the New Frontier party split in 1998 with the withdrawal of the former Kömeitö groups and its leading members were absorbed into the LDP and the Democratic Party of Japan. The former Kömeitö groups received strong backing from Söka Gakkai, a Buddist school association, and reunified as the New Kömeitö at the end of 1998. It joined a ruling coalition with the LDP in 1999.

8. According to Iida, MITI's hidden motive for introducing the 'green power fund' through its Public Utilities Department was to undermine the arguments supporting the proposed 'Green Energy Law', both in favor of the industry and to retain its policy initiative (Iida 2002: 9).

9. Even after the statewide power crisis of early 2001, Sacramento Municipal Utility District (SMUD) still offers a green power product generated from 100% renewable energy at a fixed rate of $6/month or $0.01/kW (*http://www.smud.org/green/summary.html*, 20 May 2001).

10. Japan Natural Energy Co. Ltd. (Nihon Shizen Enerugï Co. Ltd.: JNEC), established in November 2000 with funding from TEPCO and others, issues 'green energy certificates' to corporations and regional authorities in a voluntary scheme. At the end of October 2002, 27 companies and one regional government had joined the scheme. The customers consume and pay for their electricity as usual, but on top of that, they purchase a certified amount of a notional green power product in the form of 'Certification of Green Power' at 4 yen/kW. To illustrate, company A uses 3 million kW/year, which it buys from the utility company at market price. If it then buys an equivalent amount of 'green energy certificates', it is certified as 100% green powered. (The first in the country to get a 100% green power certification was the Sony tower at Shinsaibashi, Osaka.) The generated power is purchased by a regional electric power company in the power generation facility area at some price. The wind power generators receive money both from the sales of electricity and the certificates which are issued according to the amount of electricity generated by JNEC, the certifying body. With Certification of Green Power, customers can show proof of their use of green electricity. This can serve various purposes such as meeting fossil fuel saving and CO_2 emission reduction targets, obtaining an environmental ISO and improving corporate identity. Separating 'sales of environmental value added' electricity from 'sales of electric power' in general is a key feature of this system, enabling both the provision and purchase of notional 'green power' regardless of one's proximity to suitable wind generating areas or which electric power company operates in one's region. (*http://www.natural-e.co.jp/english/index.html*, 31 October 2003.)

11. Based on an interview with the originators of the FCR, including Wolf von Fabeck, managing director of Solar Promotion Association (Aachen's Solarenergie-Forderverein) on 18 March 2002 (see *http://www.sfv.de/infos/soinf171.htm*, 10 April 2002). This interview and others relating to Germany's agreement on post nuclear society (footnotes 13 and 14 below) were conducted jointly with Yüko Takubo of Fuji Tokoha University during March 2002. See Hermann Scheer (2001) for a discussion of green power schemes focusing on solar energy.

12. Some of the main points of the revised Atomic Energy Act include: the allowable lifespan of each reactor was set at 32 years from the time it

commenced operations. On this basis, a notional lifetime output for each reactor was calculated, with any remainder after decommissioning transferable to another plant. Hence an older and less efficient reactor can be shut down before its allocated 32 years has expired and its remaining output transferred to another reactor, allowing the latter to remain on the grid for longer than 32 years. The newest reactor went into commercial operation in April 1989, so nuclear power in Germany is not expected to be totally eliminated until sometime after 2021, since transfers of quotas from other reactors will enable this one to operate for more than 32 years.

Furthermore, no new reactors are to be built, and fuel reprocessing is totally prohibited from 1 July 2005. Nuclear waste is to be mainly stored in interim storage facilities to be constructed on-site or near the existing reactors. Finally, the maximum limit for a compensation payout for a serious accident was increased tenfold to 2.5 billion Euros.

13. The agreed 32 year reactor lifespan was a compromise. The industry had asked for 35 years, while the Environment Minister (Jürgen Trittin of the Greens) offered 25 years. Agreement was reached only after the government accepted the transfer of remaining power limits from decommissioned reactors to others. The Stade reactor, operating since May 1972, was permanently closed for economic reasons in November, 2003. Since its allocated production life-span was until mid-2004, it was able to transfer its outstanding production quota to other plants.

Environmentalists have criticized this agreement for favoring the nuclear industry too much. But if the government had uncompromisingly pursued the environmentalists' ideals and passed laws forcing the immediate closure of the nuclear power plants, there would have been a lengthy legal battle as the industry sued the government for violating constitutionally protected property rights and sought financial compensation. The potential costs of losing would be much too high for the government to risk and the compromise eventually would achieve the same objective, that is, phasing out the nuclear plants.

The industry also compromised in abandoning its claims for compensation. A huge compensation bill would have caused a national outrage. The compromise thus averted a potential political crisis, and gave the industry nearly two decades of operating time. In the end the compromise ensured political stability, avoided lengthy and expensive court battles, and delivered the world's first concrete program to phase out nuclear power.

14. I examined the results of the collaboration between the car makers, electric utilities and the Rocky Mountain Institute (led by Amory Lovins, the originator of the 'soft energy path' concept) (1996a).

Chapter 11

1. This chapter was originally published as a comprehensive opening article in issue number 200 of the *Japanese Sociological Review* (Japan Sociological Society 2000). This issue was a special feature entitled 'Sociological imagination towards the 21st century—the new communality and publicness'.

2. There have been numerous and extensive scholarly publications about the revival of public philosophy and theories of the public sphere or publicness in Japan in the past few years; for example the 10 volume set entitled *Public Philosophy* (*Kōkyō tetsugaku*) (Sasaki and Kim eds, 2001–2002), with

contributions from scholars in the social sciences, humanities and natural sciences, and Takatoshi Imada (2001).

3. In his account of the lawsuit filed by the Sierra Club to stop a large resort development, Stone explains that the word 'standing' in the article's title has a double meaning: standing as a witness to give evidence in court, and legal standing to qualify as a plaintiff (Stone 1972). Although the plaintiff lost, the case became famous in 1972 when US Federal Supreme Court Judge Douglas quoted and supported Stone's argument in his 'dissenting opinion'.

4. This discussion is from the Forward to the Japanese edition of 1991, and is not included in the original English edition.

5. These phrases are from the Forward to the Japanese edition of 1991, and are not included in the original English edition.

6. Nitagai (2001: 43–4) proposed a model of the public sphere that emphasizes autonomy, 'intimacy', free will and flexibility, based on his analysis of volunteer activities in the Great Hanshin-Awaji Earthquake.

7. Refer to Calhoun (1992) and Abe (1998: 170–224) for critiques of Habermas' theory of the public sphere, especially on its pessimism, and on the absence of any analysis of the private interests of the capitalist media organizations and their affects on the media's role in the public sphere.

8. Torigoe's first proposal of this concept (1995) was not clear on this point, but in a note two years later (1997: 79) he explained that 'the communal seizin in this case study exists in the balance of the tug-of-war between two legal bodies. The holder of the communal seizin always functions as a legal body'. But still, questions remain. Why can individual residents not possess this right? Can a local residents' organization, a neighborhood organization, an association or a group hold it? How should the difference in the value or content of the communal seizin claimed among any disputants be resolved?

9. I follow Fujii (1999) in adopting the term 'citizens' sector'.

10. Yoshihara describes this phenomenon as 'a rebuilding of the public space by the citizens' from the perspective of local governance theory, and considers parallels to developments in the 'theories of new civil society' (2002: 117).

Chapter 12

1. Shōji (1999: 177–8) posits conviviality as a social science concept and indicates four versions, comprising co-existence, sharing, symbiosis and sympathy.

2. According to Tetsurō Yoshimoto of the Minamata City Welfare and Life-style Department, Environmental Division on 12 July 1999. See also (Yoshimoto 1995).

3. The price of fuel for a nuclear power plant is cheaper at 1.7 yen per kWh (as estimated by METI). Thus, in spite of enormous construction costs (approximately 300–400 thousand yen per kWh), operating a nuclear power plant remains competitive with an oil fired power plant burning relatively expensive heavy oil at 3–4 yen per kWh. Clearly, the economic advantages of nuclear power increase as the price of oil increases, as it did after the oil crises of the 1970s.

4. According to a website index of European sustainable city initiatives, of the 138 recorded cases, 54 (40%) are energy related projects (*http://www.eaue.de/winuwd/* 8 June 2002), an indication of the prominence of energy in the creation of sustainable cities.

5. From discussions with Sailer on 27 July 1994 and 21 March 2002. Another discussion in March 2002 included Yüko Takubo of Fuji Tokoha University.

Chapter 13

1. What is here called the 'citizens' sector' has also been called the 'private, non profit sector', 'non profit, cooperative sector', 'communal sector', 'social sector' and 'social economy'. In English it is generally referred to as the 'non profit sector'.
2. The institutionalization of social movements is a global trend. Mitchell et al. (1992) and Terada (1998a, 1998b) examine the advantages and disadvantages of the institutionalization of environmental movements in the USA. For international research into the socio-political background and effects of social movement institutionalization, see Meyer and Tarrow eds (1998).
3. McAdam, Tarrow and others' discuss various problems with the theory of political opportunity structures (McAdam et al. eds, 1996; Tarrow 1994, and chapter 5 of this book).
4. The rankings for the seven years from 1997 to 2003 are available on the Japanese Citizens' Ombudsman Association web page (*http://www.ombudsman.jp/*, 5 June 2003).

Conclusion

1. KFEM was founded in 1993. With its 85,000 members and 47 local branches, it is the largest and the most influential NGO and the largest environmental organization in Korea (see Kim and McNeal forthcoming).
2. TEPU was founded in 1987. With about one thousand members and 10 local chapters, it is the leading environmental organization in Taiwan. It has a strong influence on the environmental policies of President Chen through its very close relationships with him and his party (see Hsiao forthcoming).
3. In Korea, Kim Dae-jung was President from 1998 to 2003 and in Taiwan Chen Shui-bian has held the office since 2000. Both had previously been leading activists and figureheads in their respective countries' pro-democracy movements.

Bibliography

Note that all Japanese language books are published in Tokyo unless otherwise specified.

Abe, Kiyoshi (1998), *Kökyöken to komyunikeishon: Hihanteki kenkyü no arata na chihei* (Communication and the public sphere: New horizons in critical research), Kyoto: Mineruba Shobö.

Abe, Yasutaka and Takehisa Awaji (1998), *Kankyöhö, dai ni han* (Environ-mental law, 2nd edition), Yühikaku.

Adachi, Shigekazu (1999), 'Chiiki kankyö undö no ishi kettei to jümin no söi: Gifu Ken X Chö no Nagaragawa kaközeki kensetsu hantaiha no jirei kara (Decision-making process in local environmental movements and consensus building among community people: A case study of the movement against the construction of the damming of the Nagara river estuary dam in X Town, Gifu Prefecture)', *Kankyö shakaigaku kenkyü* (Journal of environmental sociology), 5, pp. 152–65.

Aoi, Kazuo (1993), 'Shakaigaku (Sociology)', in Kiyomi Morioka, Tsutomu Shiobara and Yasuhei Honma (eds), *Shin shakaigaku jiten* (New encyclopedia of sociology), Yühikaku, pp. 599–602.

Aoki, Shinji (2001), 'Yüki nögyö undö no kanösei (Possibilities of the organic farming movement)', Hiroyuki Torigoe (ed.), *Köza kankyö shakaigaku, 3: Shizen kankyö to kankyö bunka* (Environmental sociology in Japan, vol. 3: The natural environment and environmental culture), Yühikaku, pp. 133–57.

Asaoka, Mie (1998), 'Shimin katsudö no yakuwari to kanösei: "Kikö Föramu" no ichinen de mietekita mono (The role and possibility of citizens' activity in Japan: Days of Kiko Forum)', *Kankyö shakaigaku kenkyü* (Journal of environmental sociology), 4, pp. 77–80.

Ayukawa, Yurika (1998), 'Kyoto kaigi ni muketa ichinen: Kokusai NGO de no keiken (The year towards the Kyoto Conference: My experience at an international environmental NGO)', *Kankyö shakaigaku kenkyü* (Journal of environmental sociology), 4, pp. 85–8.

Awaji, Takehisa (1973), 'Kögai funsö no kaiketsu höshiki to jittai (The reality of pollution disputes and methods of resolution)', in *Chüshaku kögaihö taikei, dai yon kan: Funsö shori, higaisha kyüsaihö* (Compendium of pollution law commentary, vol. 4: Dispute handling and victim relief law), Nippon Hyöronsha, pp. 1–36.

———— (1980), *Kankyöken no höri to saiban* (Environmental rights judicial trials and their legal principles), Yühikaku.

Awaji, Takehisa, Kazuhiro Ueta and Köichi Hasegawa (eds) (2001), *Kankyö seisaku kenkyü no frontier: Gakusaiteki köryü to tenbö* (Frontier of environmental policy studies: Interdisciplinary symposium and the prospect), Töyö Keizai Shimpösha.

Beck, Ulrich (1986), *Risikogesellschaft: Auf dem Weg in eine andere Moderne*,

Frankfurt/Main: Suhrkamp Verlag. Mark Ritter (tr.) (1992), *Risk Society: Towards a New Modernity*, Newbury Park: Sage Publications.

Beck, Ulrich, Anthony Giddens and Scott Lash (1994), *Reflexive Modernization: Politics, Tradition and Aesthetics in the Modern Social Order*, Cambridge: Polity Press.

Bell, Michael Mayerfield (1998), *An Invitation to Environmental Sociology*, Thousand Oaks: Pine Forge Press.

Bellah, Robert N., Richard Madsen, William M. Sullivan, Ann Swidler and Steven M. Tipton (1985), *Habits of the Heart: Individualism and Commitment in American Life*, Berkeley: University of California Press.

————— (1991), *The Good Society*: New York: Knopf.

Benford, Robert D. and David A. Snow (2000), 'Framing Processes and Social Movements: An Overview and Assessment', *Annual Review of Sociology*, 26, pp. 611–39.

Broadbent, Jeffrey (1998), *Environmental Politics in Japan: Networks of Power and Protest*, Cambridge: Cambridge University Press.

Bullard, Robert D. (1994), *Dumping in Dixie: Race, Class and Environmental Quality*, 2nd ed., Boulder: Westview Press.

Buttel, Frederick H. (1987), 'New Directions in Environmental Sociology', *Annual Review of Sociology*, 13, pp. 465–88.

Calhoun, Craig (ed.) (1992), *Habermas and the Public Sphere*, Cambridge: The MIT Press.

Callenbach, Ernest, Fritjof Capra and Sandra Marburg (1990), *EcoManagement: The Elmwood Guide to Eco-Auditing and Ecologically Conscious Management*, Berkeley: Elmwood Institute.

Catton, William R., Jr and Riley E. Dunlap (1978), 'Environmental Sociology: A New Paradigm', *The American Sociologist*, 13, pp. 41–9.

————— (1980), 'A New Ecological Paradigm for Post-Exuberant Sociology', *American Behavioral Scientist*, 24, pp. 15–47.

Cohen, Jean L. (1985), 'Strategy or Identity: New Theoretical Paradigms and Contemporary Social Movements', *Social Research*, 52 (4), pp. 663–716.

Cohen, Jean and Andrew Arato (1992), *Civil Society and Political Theory*, Cambridge: The MIT Press.

Dunlap, Riley E. (1995), 'Toward the Internationalization of Environmental Sociology: An Invitation to Japanese Scholars', Kōichi Hasegawa (tr.) (1998), 'Kankyō shakaigaku no kokusaika ni mukete: Nihon no kankyō shakaigakusha e', *Kankyō shakaigaku kenkyū* (Journal of environmental sociology), 1, pp. 73–85.

————— (2002a), 'Environmental Sociology: A Personal Perspective on Its First Quarter Century', *Organization and Environment*, 15, pp. 10–29.

————— (2002b), 'Paradigms, Theories and Environmental Sociology', R. E. Dunlap, F. H. Buttel, P. Dickens and A. Gijswit (eds), *Sociological Theory and the Environment*, Lanham: Rowman and Littlefield.

Dunlap, Riley E. and William R. Catton Jr. (1979), 'Environmental Sociology', *Annual Review of Sociology*, 5, pp. 243–73.

Dunlap, Riley E., Frederick H. Buttel, Peter Dickens and August Gijswijt (eds) (2002), *Sociological Theory and the Environment*, Lanham: Rowman and Littlefield.

Foljanty-Jost, Gesine (2000) 'Kankyö seisaku no seikö no jöken: Kankyö hogo ni okeru Nihon no senkushateki yakuwari no kyöryü to shüen (Conditions for the success of environmental policies: The rise and demise of Japan's pioneering role in environmental preservation)', Minoru Tsubogö (tr.), *Leviathan*, 27, pp. 35–48.

Fujii, Atsushi (1999), "Shimin jigyö soshiki' no shakaiteki kinö to sono jöken: Shiminteki senmonsei (The social function and requisites for 'civil enterprise organizations': Civil specialities)', Yasuo Kakurai and Kiyofumi Kawaguchi (eds), *Hieiri kyödö soshiki no keiei* (Management of nonprofit, cooperative organizations), Kyoto: Mineruba Shobö, pp. 177–206.

Funabashi, Harutoshi (1988), 'Közöteki kinchö no rensateki ten'i (The vicious chain of structual tension)', in Harutoshi Funabashi, Köichi Hasegawa, Söichi Hatanaka and Takamichi Kajita (eds), *Kösoku bunmei no chiiki mondai: Töhoku shinkansen no kensetsu funsö to shakaiteki eikyö* (Regional problems in a high speed civilization: The dispute over construction of the Töhoku bullet train line and its social impacts), Yühikaku, pp. 155–87.

———— (1990), 'Shakai seigyo no san suijun: Shinkansen kögai taisaku no nichifutsu hikaku o jirei toshite (Three levels of social control system: A case study comparing French and Japanese protection measures against pollution from the bullet train)', *Shakaigaku hyöron* (Japanese sociological review), 41 (3), pp. 73–87.

———— (1998), 'Kankyö mondai no mirai to shakai hendö: Shakai no jiko-hakaisei to jikososhikisei (Social change and the future of environmental problems: The self-organization and self-destructiveness of society)', in Harutoshi Funabashi and Nobuko Iijima (eds), Köza shakaigaku, 12: Kankyö (Sociology in Japan, vol. 12: Environment), University of Tokyo Press, pp. 191–224.

———— (2000), 'Kumamoto Minamatabyö no hassei kakudai katei ni okeru gyösei soshiki no musekininsei no mekanizumu (The mechanism of administrative institutional irresponsibility in the escalation of the outbreak of Minamata disease in Kumamoto)', in Sökan shakai kagaku yüshi (Interrelated social sciences volunteers) (eds), Weber, Durkheim, Nihonshakai: Shakaigaku no koten to gendai (Weber, Durkheim and Japanese society: The classics of sociology and contempory Japan), Häbesutosha, pp. 129–211.

———— (2001), 'Kankyö mondai no shakaigakuteki kenkyü (Sociological study on environmental problems)', in Nobuko Iijima, Hiroyuki Torigoe, Köichi Hasegawa and Harutoshi Funabashi (eds), *Köza kankyö shakaigaku, 1: Kankyö shakaigaku no shiten* (Environmental sociology in Japan, vol. 1: Perspectives of environmental sociology), Yühikaku, pp. 29–62.

Funabashi, Harutoshi, Köichi Hasegawa, Söichi Hatanaka and Harumi Katsuta (1985), *Shinkansen kögai: Kösoku bunmei no shakai mondai* (Bullet train pollution: Social problems of a high speed civilization), Yühikaku.

Funabashi, Harutoshi, Köichi Hasegawa, Söichi Hatanaka and Takamichi Kajita (1988), *Kösoku bunmei no chiiki mondai: Töhoku shinkansen no kensetsu funsö to shakaiteki eikyö* (Regional problems in a high speed civilization: The dispute over construction of the Töhoku bullet train line and its social impacts), Yühikaku.

Funabashi, Harutoshi and Akira Furukawa (eds) (1999), *Kankyō shakaigaku nyūmon: Kankyō mondai kenkyū no riron to gihō* (Introduction to environmental sociology: theories and techniques of research into environmental problems), Bunka Shobō Hakubunsha.

Funabashi, Harutoshi, Kōichi Hasegawa and Nobuko Iijima (eds) (1998), Kyodai chiiki kaihatsu no kösö to kiketsu: Mutsu Ogawara kaihatsu to kakunenryö saikuru shisetsu (Vision versus results in a large-scale industrial development project in the Mutsu-Ogawara district: a sociological study of social change and conflict in Rokkasho village), University of Tokyo Press.

Funabashi, Harutoshi and Nobuko Iijima (eds) (1998), *Köza shakaigaku, 12: Kankyö* (Sociology in Japan, vol. 12: Environment), University of Tokyo Press.

Funabashi, Harutoshi, Kazunori Kado, Yöichi Yuasa and Hiromitsu Mizusawa (2001), *'Seifu no shippai' no shakaigaku: Seibi shinkansen kensetsu to kyükokutetsu chöki saimu mondai* (The sociology of 'governmental failure': Construction of new bullet train lines and the problem of the long term debt of the former Japan National Railway), Häbesutosha.

Furukawa, Akira (1984), 'Kawa to ido to mizuumi: Kogan shüraku no dentöteki yöhaisui (River, well and lake: Traditional methods of drainage in lakeside villages)', in Hiroyuki Torigoe and Yukiko Kada (eds), *Mizu to hito no kankyöshi: Biwako hökokusho* (Environmental history of water and people: The Lake Biwa report), Ochanomizu Shobö, pp. 242–77.

Galbraith, John K. (1952), *American Capitalism: The Concept of Countervailing Power*, Boston: Houghton Mifflin.

Genshiryoku Shiryö Jöhöshitsu (Citizen's Nuclear Information Center) (eds) (2002), *Genshiryoku shimin nenkan 2002* (The nuclear energy citizens' yearbook, 2002), Nanatsu Mori Shokan.

Geschwender, James A. (1968) "Explorations in the theory of social movements and revolutions." *Social Forces*, 47 (2), pp. 127–35.

Giugni, Marco G., Doug McAdam and Charles Tilly (1998), *From Contention to Democracy*, Lanham: Rowman and Littlefield.

Gö, Hirobumi, Akio Tomishima, Katsumi Yoshida, Makoto Shimizu and Ken'ichi Miyamoto (1972), '"Zadankai" chiiki kaihatsu o sabaku: Yokkaichi kögai saiban ni tsuite ('Symposium' to adjudicate a regional development project: On the Yokkaichi pollution lawsuits)', Kögai kenkyü (Research on environmental disruption), 2 (2), pp. 38–51.

Gurr, Ted R. (1970) *Why Men Rebel*, Princeton: Princeton University Press.

Habermas, Jürgen (1981), 'New Social Movements', *Telos*, 49, pp. 33–7.

———— (1990), *Strukturwandel der Öffentlichkeit: Untersuchungen zu einer Katergorie der bürgerlichen Gesellschaft*, Frankfurt/Main: Suhrkamp Verlag.

Hanada, Tatsurö (1996), *Kökyöken to iu na no shakai kükan: Kökyöken, media, shimin shakai* (A social space named the public sphere: The public sphere, media and civil society), Bokutakusha.

Hannigan, John A. (1995), *Environmental Sociology: A Social Constructionist Perspective*, New York: Routledge.

Harashina, Sachihiko (ed.) (2000), *Kankyö asesumento, kaiteiban* (Environmental impact assessment, revised edition), Hösö Daigaku Kyöiku Shinkökai.

Hasegawa, Kōichi (1985a), 'Shakai undö no seiji shakaigaku: Shigen döinron no igi to kadai (The political sociology of social movements: The significance of and issues in resources mobilization theory), *Shisö* (Thought), 737, Tokushü: Atarashii shakai undö (Special issue: New social movements), pp. 126–57.

―――― (1985b), 'Kögai soshö to jümin undö (Anti-pollution lawsuits and residents' movements), in Funabashi, Harutoshi, Köichi Hasegawa, Söichi Hatanaka and Harumi Katsuta, *Shinkansen kögai: Kösoku bunmei no shakai mondai* (Bullet train pollution: Social problems of a high speed civilization), Yühikaku, pp. 207–35.

―――― (1989), '"Gendaigata soshö" no shakai undöronteki kösatsu: Shigen döin katei to shite no saiban katei (Examining 'contemporary lawsuits' from the perspective of social movements: The judicial trial as a resource mobilizing process)', *Höritsu jihö* (Law review), 61 (12), pp. 65–71.

―――― (1990), 'Shigen döinron to "atarashii shakai undö" ron (Resource mobilization theory and 'new social movement' theory)', in Shakai undöron kenkyükai (Research group on social movement theories) (eds), *Shakai undöron no tögö o mezashite* (Toward the study of social movement theory), Seibundö, pp. 3–28.

―――― (1991a), 'Chihö kyoten toshi ni okeru hangenshiryoku undö no undö katei (Movement processes of antinuclear energy movements in regional, urban bases)', in Akira Takahashi (ed.), *Toshi keikaku to toshi shakai undö no sögöteki kenkyü* (General research on urban planning and urban social movements), 1989 Nendo kagaku kenkyühi hojokin kenkyü seika hökokusho (1989 Academic year Mombusho research grant, research report), Shizuoka Kenritsu Daigaku, pp. 7–47.

―――― (1991b), 'Hangenshiryoku undö ni okeru josei no ichi: Posuto Cherunobuiri no "atarashii shakai undö" (The status and role of women in post Chernobyl anti-nuclear energy movements),' *Leviathan*, 8, pp. 41–58.

―――― (1993a), 'Shakai undö: Fuman to döin no dainamizumu (Social movements: The dynamism of discontent and mobilization process)', in Takamichi Kajita and Nobuyoshi Kurita (eds), *Kïwädo: Shakaigaku* (Learning from keywords of sociology), Kawashima Shoten, pp. 147–63.

―――― (1993b), 'Kankyö mondai to shakai undö (Environmental problems and social movements)', in Nobuko Iijima (ed.), *Kankyö shakaigaku* (Environmental sociology), Yühikaku, pp. 101–22.

―――― (1993c), 'Shakai undö toshite no kenpö sosyö (Constitutional litigation as a strategy of protest movements)', *Höritsu jihö* (Law review), 65 (11), pp. 58–62.

―――― (1996a), 'Toshi kükan ni okeru keikaku to undö (Planning and social movements in the urban space)', in Naoki Yoshihara (ed.), *Nijüisseiki no toshi shakaigaku, 5: Toshi kükan no kösöryoku* (Urban sociology for the twenty-first century, vol. 5: The imagination from the urban space), Keisö Shobö, pp. 125–63.

―――― (1996b), 'NPO: Datsugenshiryoku seisaku no pätonä (Study on NPOs: A transformation process of energy policy)', *Sekai* (World), 643, pp. 244–54.

―――― (1996c), *Datsugenshiryoku shakai no sentaku: Shin enerugï kakumei no jidai* (A choice for a post-nuclear society: The age of new energy revolution), Shin'yösha.

―――― (1996d), 'Kankyö shakaigaku to seisaku kenkyü (Environmental sociology and policy studies)', in Kankyö Keizai Seisaku Gakkai (The Society for Environmental Economics and Policy Studies) (ed.), *Kankyö keizai seisaku kenkyü no furontia* (Frontiers in environmental economics and policy research), Töyö Keizai Shimpösha, pp. 134–41.

———— (1997), 'Chikyü ondanka mondai no kashika no tameni (Bringing global warming into the open)', *Sekai* (World), 643, pp. 93–102.

———— (1998a), 'Kakunen hantai undö no közö to tokushitsu (The structure and main characteristics of the anti-nuclear fuel cycle movement)', in Harutoshi Funabashi, Köichi Hasegawa and Nobuko Iijima (eds), *Kyodai chiiki kaihatsu no kösö to kiketsu: Mutsu Ogawara kaihatsu to kakunenryö saikuru shisetsu* (Vision versus results in a large-scale industrial development project in the Mutsu-Ogawara district: a sociological study of social change and conflict in Rokkasho village), University of Tokyo Press, pp. 249–70.

———— (1998b), 'Kyoto kaigi no genba kara (From the COP3 Kyoto conference)', *Shosai no mado* (Library window), 472, pp. 10–13.

———— (1998c), 'Yomigaetta Sendai no sora: supaiku taiya mondai kara manabu (The clearing Sendai sky: Learning from studded tyre pollution)', *Shosai no mado* (Library window), 473, pp. 10–15.

———— (1998d), 'Paburikku to "öyake" no aida: NPO höan no seiritsu o ukete (Between definitions of 'public': After the enactment of the NPO law)', *Shosai no mado* (Library window), 475, pp. 10–14.

———— (1998e), 'Shinkansen kögai: Moto kokutetsu gishichö Shima Hideo Shi no közai (Pollution from the bullet train: Advantages and disadvantages of a former chief engineer of the Japan National Railway, the late Hideo Shima)', *Shosai no mado* (Library window), 476, pp. 44–7.

———— (1998f), 'Jümin töhyö ga toikakerumono (Social implications posed by the public referendum)', *Shosai no mado* (Library window), 477, pp. 38–42.

———— (1998g), 'Füsha ga hiraku mirai: Yamagata Ken Tachikawa Machi no torikumi (The future opened by a wind generator: The efforts of Tachikawa Town, Yamagata Prefecture)', *Shosai no mado* (Library window), 480, pp. 33–8.

———— (1999a), 'Genshiryoku hatsuden o meguru Nihon no seiji, keizai, shakai (Japanese politics, economics and society on nuclear energy)', in Yoshikazu Sakamoto (ed.), *Kaku to ningen I: Kaku to taiketsu suru nijü seiki* (Confronting nuclearism vol. 1: The 20th century world in crisis), Iwanami Shoten, pp. 281–337.

———— (1999b), '"Rokkasho Mura" to "Maki Machi" no aida: Genshiryoku shisetsu o meguru shakai undö to chiiki shakai (Social movements and local communities located near nuclear facilities: Thinking on the Maki Machi and Rokkasho cases)', *Shakaigaku Nenpö* (Annual reports of the Tohoku sociological society), 28, pp. 53–76.

———— (1999c), 'Global Climate Change and Japanese Nuclear Policy', *International Journal of Japanese Sociology*, 8, pp. 183–97.

———— (2000a), 'Kyödösei to kökyösei no gendaiteki isö (The sociological aspects of communality and publicness)', *Shakaigaku hyöron* (Japanese sociological review), 50 (4), pp. 436–50.

———— (2000b), 'Downstream e no manazashi: Kankyö mondai to kankyö shakaigaku (An eye to downstream: Environmental problems and environmental sociology)', *Jökyö* (Situation), 11 (7), Hachi gatsu gö bessatsu: Gendai shakaigaku no saizensen 3 (August edition supplement: The frontier of current sociology, 3), pp. 234–45.

—————— (2000c), 'Shimin ga kankyō borantia ni naru kanōsei (The sociological condition of citizen's activities as environmental volunteers)', in Hiroyuki Torigoe (ed.), *Kankyō borantia, NPO no shakaigaku* (Sociology of environmental volunteers and NPOs), Shin'yōsha, pp. 177–92.

—————— (2000d), 'Hōshasei haikibutsu mondai to sangyō haikibutsu mondai (Radioactive and industrial waste problems from a sociological perspective)', *Kankyō shakaigaku kenkyū* (Journal of environmental sociology), 6, pp. 66–82.

—————— (2001a), 'Kankyō undō to kankyō kenkyū no tenkai (Environmental movements and the development of environmental studies)', in Nobuko Iijima, Hiroyuki Torigoe, Kōichi Hasegawa and Harutoshi Funabashi (eds), *Kōza kankyō shakaigaku, 1: Kankyō shakaigaku no shiten* (Environmental sociology in Japan, vol. 1: Perspectives of environmental sociology), Yūhikaku, pp. 89–116.

—————— (2001b), 'Sunabukuro ga katarukoto: NGO to seifu no aida (What the sandbags are talking: Between NGOs and the government)', *Kankyō to kōgai* (Research on environmental disruption), 30 (4), p. 53.

—————— (2001c), 'Kankyō undō to kankyō seisaku (Environmental movements and environmental policy)', in Kōichi Hasegawa (ed.), *Kōza kankyō shakaigaku, 4: Kankyō undō to seisaku no dainamizumu* (Environmental sociology in Japan, vol. 4: Dynamism of environmental movements and policy), Yūhikaku, pp. 1–34.

—————— (2002a), 'Kankyō shakaigaku to kankyō hōgaku (Environmental sociology and the study of environmental law)', in Tadashi Otsuka and Yoshinobu Kitamura (eds), *Kankyō hōgaku no chyōsen: Awaji Takehisa Kyōju and Abe Yasutaka Kyōju kanreki kinen* (Toward the study of environmental law: In commemoration of the sixtieth birthdays of Professor Takehisa Awaji and Professor Yasutaka Abe), Nihon Hyōronsha, pp. 341–54.

—————— (2002b), 'NPO to atarashii kōkyōsei (NPOs and the new public sphere)', in Takeshi Sasaki and Tae-Chang Kim (eds), *Kōkyō tetsugaku, 7: Chūkan shūdan ga hiraku kōkyōsei* (Public philosophy, vol. 7: Intermediary organizations and publicness), University of Tokyo Press, pp. 1–17.

—————— (2003), 'Kankyō shakaigaku to toshi shakaigaku no aida (Toward a new relationship between environmental sociology and urban sociology)', *Nihon toshi shakaigakkai nenpō* (The annals of the Japan Association for Urban Sociology), 21, pp. 23–38.

Hasegawa, Kōichi (ed.) (2001), *Kōza kankyō shakaigaku, 4: Kankyō undō to seisaku no dainamizumu* (Environmental sociology in Japan, vol. 4: Dynamism of environmental movements and policy), Yūhikaku.

Hasegawa, Kōichi and Harutoshi Funabashi (1998), 'Daikibo kaihatsu projekuto to chiiki mondai (Large scale development projects and regional problems)', in Harutoshi Funabashi, Kōichi Hasegawa, Sōichi Hatanaka and Takamichi Kajita (1988), *Kōsoku bunmei no chiiki mondai: Tōhoku shinkansen no kensetsu funsō to shakaiteki eikyō* (Regional problems in a high speed civilization: The dispute over construction of the Tōhoku bullet train line and its social impacts), Yūhikaku, pp. 1–42.

Hasegawa, Kōichi and Sōichi Hatanaka (1985), 'Jūmin undō to chiiki shakai (Residents' movement and community)', in Harutoshi Funabashi, Kōichi Hasegawa, Sōichi Hatanaka and Takamichi Kajita (1985), *Shinkansen kōgai:*

Kōsoku bunmei no shakai mondai (Bullet train pollution: Social problems of a high speed civilization), Yūhikaku, pp. 175–203.

Hasegawa, Kōichi, Hiroyuki Torigoe and Harutoshi Funabashi (2002), ''Kōza kankyō shakaigaku (zen go kan)' no kikaku, hensyü ni atatte (On the planning and editing of five volumes of 'Environmental sociology in Japan')', in Iijima Nobuko Sensei Kinen Kankō Iinkai (Nobuko Iijima Memorial Publication Editorial Committee) (eds), *Kankyō mondai to tomoni: Iijima Nobuko sensei tsuitō bunshü* (Together with environmental problems: The Nobuko Iijima memorial anthology), pp. 211–3.

Hashizume, Daisaburo (2000), 'Kōkyōsei towa nanika (On publicness)', *Shakaigaku hyōron* (Japanese sociological review), 50 (4), pp. 451–63.

Hasumi, Otohiko (1965), 'Kaihatsu gyōsei to jūmin undō (Development administrations and residents' movements)', in Tadashi Fukutake (ed.), *Chiiki kaihatsu no kōsō to genjitsu, 3* (Planning and reality in regional development projects, vol. 3), pp. 63–107.

Hidaka, Rokurō (1973), 'Shimin to shimin undō (Citizens and citizens' movements)', *Iwanami kōza: Gendai toshi seisaku, 2, shimin sanka* (Iwanami lectures: Modern urban policy, vol. 2: Citizen participation), Iwanami Shoten, pp. 39–60.

Hirose, Takashi (1987), *Kiken na hanashi: Cherunobuiri to Nihon no unmei* (On the risk of nuclear power stations: Chernobyl and Japan's fate), Hachigatsu Shokan.

Hirsch, Joachim (1998), 'The crisis of Fordism: Transformations of the 'Keynesian' security state and new social movements', *Research in Social Movements, Conflicts and Change,* 10, pp. 43–55.

Hokkaidō Gurïnfando (Hokkaidō Green Fund) (1999), *Gurïn denryoku: Shiminhatsu no shizen enerugï seisaku* (Green electricity: Citizen initiated renewable energy policy), Komonzu.

Hojō, Sachiko (1998), 'Chiiki NGO to Kyoto kaigi (Local NGO network and the COP3 Kyoto conference)', *Kankyō shakaigaku kenkyü* (Journal of environmental sociology), 4, pp. 89–92.

Hongō, Masatake (2002), 'Shakai undōron ni okeru "fureimingu" no rironteki ichi (The theoretical position of 'framing' in social movement theory),' *Shakaigaku kenkyü* (The study of sociology), 71, pp. 215–30.

Horikawa, Saburō (1998), 'Rekishiteki kankyō hozon to chiiki saisei (Historical environmental conservation and regional regeneration)', in Harutoshi Funabashi, Nobuko Iijima (eds), Kōza shakaigaku, 12: Kankyō (Sociology in Japan, vol. 12: Environment), University of Tokyo Press, pp. 103–32.

————— (1999), 'Sengo Nihon no shakaigakuteki kankyō mondai kenkyü no kiseki: Kankyō shakaigaku no seidoka to kongo no kadai (The rise and institutionalization of environmental sociology: An overview and assessment of the Japanese experience)', *Kankyō shakaigaku kenkyü* (Journal of environmental sociology), 5, pp. 211–23.

Hōritsu Jihō (1973), 'Osaka kükō saiban (Osaka airport judicial trial)', *Hōritsu jihō* (Law Review), 45 (13), Jüichigatsu gō rinji zōkan: Kōgai saiban, dai yon shü (November edition special supplement: Antipollution lawsuits, no. 4).

Hosouchi, Nobutaka (1999), *Komyunitii bijinesu* (Community business), Chüō Daigaku Shuppanbu.

Hoshikawa, Takeshi (1994), 'Jiritsuteki kōkyōsei e no kōzō tenkan ni mukete:

Shimin shakai no kiban toshite no media, nettowäkingu no kanösei (Towards a structural transformation for an autonomous public sphere: The possibilites of media and networking as bases for civil society)', *Shakaigaku hyöron* (Japanese sociological review), 45 (3), pp. 332–45.

Humphrey, Craig R. and Frederick H. Buttel (1982), *Environment, Energy and Society*, Belmont: Wadsworth.

Humphrey, Craig R., Tammy L. Lewis and Frederick H. Buttel (2001), *Environment, Energy and Society: A New Synthesis*, Belmont: Wadsworth.

Iida, Tetsunari (1999), 'Gurïn denryoku seido no tenkai (Development of green electricity systems)', *Kankyö to kögai* (Research on environmental disruption), 28 (4), pp. 31–7.

——— (2000), *Hokuö no enerugï demokurashï* (Energy democracy in Northern Europe), Shinhyöron.

——— (2002), 'Yugamerareta "shizen enerugï sokushin hö": Nihon no enerugï seisaku kettei purosesu no jissö to kadai (The twisted 'green energy law': The issues and reality of the decision making process of Japanese energy policy)', *Kankyö shakaigaku kenkyü* (Journal of environmental sociology), 8, pp. 5–23.

Iijima, Nobuko (1970a), 'Kögai hantai undö to kögai saiban (Antipollution movements and antipollution judicial trials)', *Jümin to jichi* (Residents and self government), March edition, pp. 44–51.

——— (1970b), 'Sangyö kögai to jümin undö (Industrial pollution and the community residents' movement: The case of the Minamata disease)', *Shakaigaku hyöron* (Japanese sociological review), 21 (1), pp. 25–45.

——— (1993), *Kankyö mondai to higaisha undö: kaiteiban* (Environmental problems and victims' movements: Revised edition), Gakubunsha.

——— (1993), 'Kankyö mondai no shakaigakuteki kenkyü: sono kiseki to kongo no tenbö (Sociological study of environmental problems: The history and the prospect)', in Nobuko Iijima (ed.), *Kankyö shakaigaku* (Environmental sociology), Yühikaku, pp. 213–32.

——— (1998a), 'Josei no kankyö ködö to Aomori Ken no hankaihatsu hankakunen undö (Womens' environmental activities: Anti-industrial development project and antinuclear movements in Aomori prefecture)', in Harutoshi Funabashi, Köichi Hasegawa and Nobuko Iijima (eds), *Kyodai chiiki kaihatsu no kösö to kiketsu: Mutsu Ogawara kaihatsu to kakunenryö saikuru shisetsu* (Vision versus results in a large-scale industrial development project in the Mutsu-Ogawara district: A sociological study of social change and conflict in Rokkasho village), University of Tokyo Press, pp. 271–99.

——— (1998b), 'Söron: Kankyö mondai no rekishi to kankyö shakaigaku (Review of the history of environmental problems and environmental sociology),' in Harutoshi Funabashi and Nobuko Iijima (eds), *Köza shakaigaku, 12: Kankyö* (Sociology in Japan, vol. 12: Environment), University of Tokyo Press, pp. 1–42.

——— (2000), Kankyö mondai no shakaishi (A social history of environmental problems), Yühikaku.

——— (2001), 'Chikyü kibo no kankyö mondai to shakaigakuteki kenkyü (Global scale environmental problems and sociological study)', in Nobuko Iijima (ed.), *Köza kankyö shakaigaku, 5: Ajia to sekai: Chiiki shakai kara no shiten* (Environmental sociology in Japan, vol. 5: Asian and global environmental problems: Sociological studies of local origins), Yühikaku, pp. 1–32.

——— (2002), *Iijima Nobuko kenkyü kyöiku shiryöshü* (Collected works of Iijima Nobuko), Iijima Nobuko Sensei Kinen Kankö Iinkai.

Iijima, Nobuko (ed.) (1993), *Kankyö shakaigaku* (Environmental sociology), Yühikaku.

Iijima, Nobuko (ed.) (2001), *Ajia shakai ni okeru chiiki kaihatsu to kankyö mondai ni kansuru kankyö shakaigakuteki kenkyü* (Environmental sociological studies on regional development projects and environmental problems in Asian societies), 1998–2000 Nendo kagaku kenkyühi hojokin kenkyü seika hökokusho (The 1998–2000 academic year Mombusho research grant, research report), Tokyo Metropolitan University.

Iijima Nobuko Sensei Kinen Kankö Iinkai (Nobuko Iijima Memorial Publication Editorial Committee) (eds) (2002), *Kankyö mondai to tomoni: Iijima Nobuko sensei tsuitö bunshü* (Environmental issues: Essays in memory of Iijima Nobuko).

Ikeda, Kanji (1995), 'Kankyö shakaigaku no shoyüronteki päsupekutibu: "Guröbaru komonzu no higeki" o koete (Property as a perspective of environmental sociology),' *Kankyö shakaigaku kenkyü* (Journal of environmental sociology), 1, pp. 21–37.

——— (2001) 'Chikyü ondanka böshi seisaku to kankyö shakaigaku no kadai: Poritikkusu kara gabanansu e (Possible contributions of environmental sociology to the progress of global climate protection policy: From international policies to global governance),' *Kankyö shakaigaku kenkyü* (Journal of environmental sociology), 7, pp. 5–23.

Iken Kökoku Zendö Renrakukai (All Hokkaidö Opinion Advertising Coalition) (1988), *Irranaissho! Genshiryoku hatsuden hökokusho* (We don't need it! Report on nuclear power generation).

Imada, Takatoshi (2001), *Imi no bunmeigaku josetsu: Sono saki no kindai* (A discourse on the civilization of meaning: Modernity and beyond, University of Tokyo Press.

Imai, Hajime (2000), *Jümin töhyö: Kankaku minshushugi o koete* (Public referendums: Beyond a spectator democracy), Iwanami Shoten.

Imai, Hajime (ed.) (1997), *Jümin töhyö: Nijü seikimatsu ni mebaeta Nihon no shin rüru* (Public referendums: Newly formed rules in Japan at the end of the twentieth century), Nihon Keizai Shimbunsha.

Imaizumi, Mineko (2001), *Furaiburugu kankyö repöto* (The Freiburg environmental report), Chüö Höki Shuppan.

Independent Sector (2001), *The New Nonprofit Almanac in Brief: Facts and Figures on the Independent Sector 2001, http://www.independentsector.org/PDFs/inbrief.pdf*, 31 October 2002.

Itö, Mamoru (ed.) (2001), *Chihö ni okeru atarashii shakaiteki kankei no keisei no bunpö no kaimei ni mukete: Niigata Ken Maki Machi 'Jümin töhyö' jitsugen no shakaiteki shojö ken no kaimei kara* (Towards analysis of a new grammar in the formation of new social relations in community: From the study of social conditions of public referendum in Maki Town, Niigata Prefecture), 1999–2001 Nendo kagaku kenkyühi hojokin kenkyü seika hökokusho (The 1999–2001 academic year Mombusho research grant, research report), Waseda University.

Kada, Yukiko (1995), *Seikatsu sekai no kankyögaku: Biwako kara no messeiji* (Environmental studies in community: The message from Lake Biwa), Nösangyoson Bunka Kyökai.

———— (2002), *Kankyö shakaigaku* (Environmental sociology), Iwanami Shoten.

Kajita, Takamichi (1979), 'Funsö no shakaigaku: Juekiken to jukuken (The sociology of conflict: Benefit zone and victimized zone)', *Keizai hyöron* (Economics review), May 1979, pp. 101–20. Reprinted (1988), in *Tekunokurashï to shakai undö: Taiköteki söhosei no shakaigaku* (Technocracy and social movements: The sociology of conflicting complementarity), University of Tokyo Press, pp. 3–30.

Kaku, Kensuke (1996), 'Gendai Nihon no seijigaku to kankyö mondai (Political science and environmental problems in Japan),' *Kankyö shakaigaku kenkyü* (Journal of environmental sociology), 2, pp. 148–55.

Kaku, Kensuke and Hitoshi Maruyama (eds) (1997), *Kankyö seiji e no shiten* (A perspective on environmental politics)', Shinzansha.

Kamata, Satoshi (1991), *Rokkasho Mura no kirokoku (jö/ge)* (History of the Rokkasho Village (1, 2)), Iwanami Shoten.

Kaneko, Isamu (1982), 'Komyuniti no shakai keikaku (Community social planning)', in Michihiro Okuda, Wataru Ömori, Noboru Ochi, Isamu Kaneko and Takamichi Kajita, *Komyuniti no shakai sekkei: Atarashii 'machizukuri' no shisö* (Community social design: Thoughts on new 'town building'), Yühikaku, pp. 179–221.

Kansho, Taeko (1987), *Mada, maniau no nara* (Not yet, only if we still have time), Chiyüsha.

Katagiri, Shinji (1995), *Shakai undö no chühan'i riron: Shigen döinron kara no tenkai* (Middle-range theory of social movements: Development from resource mobilization theory), University of Tokyo Press.

Katö, Hisatake (ed.) (1998), *Kankyö to rinri: Shizen to ningen no kyösei o motomete* (Environment and ethics: Seeking a symbiosis of nature and human), Yühikaku.

Kimura, Kunihiro (2002), *Daishüdan no jiremma: Syügö köi to syüdan kibo no süri* (The large number dilemma: Mathematics of collective action and group size), Kyoto: Mineruba Shobö.

Kimura, Yasuo (1982), *Osaka kükö kögai soshö yowa (1-4/kan): Osaka kükö kögai soshö genkoku bengodan repöto kara* (The memoir of the Osaka International Airport pollution lawsuit 1-4: From the legal counsel report on the Osaka International Airport pollution lawsuit)', *Hanrei jihö* (Precedent review), 1026, pp. 25–9; 1028, pp. 27–31; 1029, pp. 21–4; 1031, pp. 23–6.

Kimura, Yasuo and Kazumasa Kuboi (1978), 'Wagakuni ni okeru shüdan soshö no jitsujö to kadai: Osaka kokusai kükö kögai soshö no keiken o töshite (Issues and realities in class action lawsuits in Japan: Through the history of the Osaka International Airport pollution lawsuit)', *Juristo* (Jurist), 672, pp. 60–7.

Kitschelt, Herbert P. (1986), 'Political Opportunity Structures and Political Protest: Anti-Nuclear Movements in Four Democracies', *British Journal of Political Science*, 16, pp. 57–85.

Kokumin Seikatsu Sentä (National Consumer Affairs Center of Japan) (eds)

(1981), *Nihon no yüki nögyö undö* (The organic farming movement in Japan), Nippon Keizai Hyöronsha.

Kobe Toshi Mondai Kenkyüjo (Kobe Institute of Urban Research) (2002), *Chiiki o sasae kasseika suru komyuniti bijinesu no kadai do aratana hökösei* (New orientations and issues of community businesses assisting and activating community), NIRA Kenkyü Hökokusho.

Kösaka, Kenji (2000), 'Midoruman no susume: "yaku ni tatsu" shakaigaku nöto (1) (An encouragement to be a middleman: A note on 'useful'(1))', *Kansai Gakuin Daigaku Shakaigakubu kiyö* (Kansai Gakuin University School of Sociology journal), 87, pp. 197–206.

Köseishö Jinkö Mondai Kenkyüjo (Ministry of Health and Welfare, National Institute of Population) (eds) (1994), *Jinkö tökei shiryöshü* (Population statistics of Japan).

Kriesi, Hanspeter, Ruud Koopmans, Jan Willem Duyvendak and Marco G. Giugni (1995), *The Politics of New Social Movements in Western Europe: A Comparative Analysis,* Minneapolis: University of Minnesota Press.

Kurasawa, Susumu (1977), 'Toshiteki seikatsu yöshikiron josetsu (Thinking on urban ways of life)', in Eiichi Isomura (ed.) *Gendai toshi no shakaigaku* (Contemporary urban sociology), Kashima Shuppankai, pp. 19–29.

Kurita, Nobuyoshi (1989), 'Seiji sedai to kögi katsudö: Gekitotsu seiji no jidai ni okeru sedai kösei to seijiteki shakaika (The political generation and protest activity: Generation unit and political socialization in an age of confrontation politics)', *Shakaigaku hyöron* (Japanese sociological review), 39 (4), pp. 374–91.

Li, Yanyan (1999), 'Nihon ni okeru NPO sapöto purogramu no genjö to kadai (NPO support programs in Japan: Their present conditions and problems)', *Shakaigaku nenpö* (Annual reports of the Tohoku sociological society), 28, pp. 99–122.

——— (2002), *Borantarï katsudö no seiritsu to tenkai: Nihon to Chügoku ni okeru borantarï sekutä no ronri to kanösei* (The establishment and development of voluntary activities: The logic and possibilities of the voluntary sector in Japan and China), Kyoto: Mineruba Shobö.

Lipnack, Jessica and Jeffrey Stamps (1982), *Networking,* New York: Ron Bernstein Agency.

Lipsky, Michael (1968), 'Protest as a Political Resource,' *American Political Science Review,* 62, pp. 1144–58.

Lovins, Amory B. (1977), *Soft Energy Paths: Toward a Durable Peace,* San Francisco: The Friends of the Earth International.

Maruyama, Hisashi (1988), 'Hangenpatsu o meguru nyü ueibu towa nanika (What is the new wave of anti-nuclear energy movements?)', *Shakai undö* (Social Movements), 103, pp. 60–4.

Masamura, Kimihiro (1989), *Fukushi shakairon* (On welfare society), Söbunsha.

Masugata, Toshiko (1995), 'Yöki nögyö undö no tenkai to kankyö shakaigaku (Environmental sociology and the development of organic farming movements)', *Kankyö shakaigaku kenkyü* (Journal of environmental sociology), 1, pp. 38–51.

Matsubara, Akira (1999), ''NPO hö' ni itaru haikei to rippö katei (The background and legislative process for arriving at the 'NPO law')', in Yöichi Nakamura and Nihon NPO Sentä (Japan NPO Center) (eds), *Nihon no NPO, Nisen* (Japanese NPOs, 2000), Nippon Hyöronsha, pp. 51–63.

Matsubara, Haruo and Kamon Nitagai (eds) (1976), *Jümin undö no ronri: Undö no tenkai katei, kadai to tenbö* (The logic of residents' movements: Development processes, issues and prospects), Gakuyö Shobö.

Matsumoto, Yasuko (1998), 'Giteisho köshö o ugokashita kokusai NGO (International environmental NGOs which influenced the outcome of the Kyoto protocol's intergovernmental negotiation)', *Kankyö shakaigaku kenkyü* (Journal of environmental sociology), 4, pp. 81–4.

———— (2001), 'Kokusai kankyö NGO to kokusai kankyö kyötei (International environmental NGOs and the international environmental agreement)', in Köichi Hasegawa (ed.), *Köza kankyö shakaigaku, 4: Kankyö undö to seisaku no dainamizumu* (Environmental sociology in Japan, vol. 4: Dynamism of environmental movements and policy), Yühikaku, pp. 179–210.

Matsumura, Kazunori and Shinji Aoki (eds) (1991), *Yüki nögyö undö no chiikiteki tenkai: Yamagata Ken Takahata Machi no jissen kara* (The regional development of the organic farming movement: The practical example of Takahata Town, Yamagata Prefecture), Ie no Hikari Kyökai.

Matsushita, Ryüichi (1980), *Buzen kankyö ken saiban* (The Buzen environmental rights judicial trial), Nihon Hyöronsha.

Matsuura, Kaoru (1980), 'Minji soshö ni yoru shinkansen kögai funsö kaiketsu to sono genkai (Resolutions achieved by civil litigation to the bullet train pollution dispute and their limitations)', *Höritsu jihö* (Law review), 52 (11), pp. 20–2.

Matsuura, Satoko (ed.) (1999), *Soshite, higata wa nokotta: Intänetto to NPO* (Then mudflats remained: The role of the internet and NPOs in protest activities), Riberuta Shuppan.

McAdam, Doug (1998), *Freedom Summer*, New York: Oxford University Press.

McAdam, Doug (1996), 'Conceptual Origins, Current Problems, Future Directions', in Doug McAdam, John D. McCarthy and Mayer N. Zald (eds), *Comparitive Perspectives on Social Movements: Political Opportunities, Mobilizing Structures and Cultural Framings*, Cambridge: Cambridge University Press, pp. 23–40.

McAdam, Doug, John D. McCarthy and Mayer N. Zald (eds), (1996), *Comparative Perspectives on Social Movements: Political Opportunities, Mobilizing Structures and Cultural Framings*, Cambridge: Cambridge University Press.

McAdam, Doug, Sidney Tarrow and Charles Tilly (2001), *Dynamics of Contention*, Cambridge: Cambridge University Press.

McCarthy, John D. and Mayer N. Zald (1977), 'Resource Mobilization and Social Movements: A Partial Theory', *American Journal of Sociology*, 82 (6) pp. 1212–41.

Meadows, Donella H., Dennis L. Meadows, Jørgen Randers and William W. Behrens III (1972), *The Limits to Growth*, New York: Universe Books.

Melucci, Alberto (1984), 'An End to Social Movements? Introductory Paper to

the Sessions on 'New Movements and Change in Organizational Forms'',
Social Science Information, 23-4/5, pp. 819–35.
————— (1989), *Nomads of the Present: Social Movements and Individual Needs
in Contemporary Society*, Philadelphia: Temple University Press.
————— (1996), *Challenging Codes: Collective Action in the Information Age*,
Cambridge: Cambridge University Press.
Meyer, David S. and Sidney Tarrow (eds), (1998), *The Social Movement Society:
Contentious Politics for a New Century*, Lanham: Rowman and Littlefield.
Mitchell, Robert C., Angela G. Mertig and Riley E. Dunlap (1992), 'Twenty
Years of Environmental Mobilization: Trends Among National Environ-
mental Organizations', in Riley E. Dunlap and Angela G. Mertig (eds),
American Environmentalism: The U.S. Environmental Movement, 1970–1990,
Philadelphia: Taylor and Francis, pp. 11–26.
Miyamoto, Ken'ichi (1971), 'Jümin undö no riron to rekishi (Theory and history
of residents' movements)', in Ken'ichi Miyamoto and Akira Endö (eds), *Köza
gendai Nihon no toshi mondai, 8: toshi mondai to jümin undö* (Contemporary
Japanese urban problems, vol. 8: Urban problems and residents' movements),
Shöbunsha, pp. 2–69.
————— (1982), 'Shakai shihonron no kon'nichiteki igi (The new meaning
of social capital theory)', in Ken'ichi Miyamoto and Akira Yamada (eds),
Kökyö jigyö to gendai shihon shugi (Public works and modern capitalism),
Kakiuchi Shuppan, pp. 13–53.
————— (1989), *Kankyö keizaigaku* (Environmental economics), Iwanami
Shoten.
————— (1999), '"Kankyö no seiki" no kökyö seisaku (Public policy for the
'century of the environment')', *Kankyö to kögai* (Research on environmental
disruption), 28 (3), pp. 2–7.
Miyajima, Takashi (ed.) (2003), *Iwanami Lexique of Sociology*, Iwanami shoten.
Miyauchi, Taisuke (2001), 'Komonzu no shakaigaku: Shizen kankyö no shoyü,
riyö, kanri o megutte (Sociology of the commons: Ownership, utilization
and management of the natural environment)', in Hiroyuki Torigoe (ed.),
Köza kankyö shakaigaku, 3: Shizen kankyö to kankyö bunka (Environmental
sociology in Japan, vol. 3: The natural environment and environmental
culture), Yühikaku, pp. 25–46.
Miyazawa, Setsuo (1989), 'Kenri no keisei to shakai undö: Kyödö kenkyö
(Social movements and the formation of legal rights)', *Höritsu jihö* (Law
review), 61 (12), pp. 58–9.
————— (1994), *Hö katei no riariti: Höshakaigaku fïrudo nöto* (The reality of
legal processes: Legal sociology field notes), Shinzansha.
Mizutani, Yöichi (ed.) (2000), *Nisen jönen chikyö ondanka böshi shinario* (A
year 2010 scenario for the prevention of global warming), Jikkyö Shuppan.
Mol, Arthur, P. J. and Gert Spaargaren, (2000), 'Ecological Modernization
Theory in Debate: A Review', *Environmental Politics*, 9 (1), pp. 17–49.
Moore, Wilbert E, (1963), *Social Change*, New York: Prentice-Hall.
Mori, Mototaka (1996), *Zushi no shimin undö: Ikego Beigun jütaku kensetsu
hantai undö to minshu shugi no kenkyü* (Zushi residents' movement: A
study of democracy and the protest movement against construction of a US
military housing complex at Ikego forests), Ochanomizu Shobö.
Morioka, Kiyoshi (1993), 'Toshiteki seikatsu yöshiki (Urban ways of life)',

in Kiyoshi Morioka, Tsutomu Shiobara and Yasuhei Honma (eds), *Shin shakaigaku jiten* (New encyclopedia of sociology), Yōhikaku, pp. 1094–5.

Morotomi, Tōru (2002), 'Chiiki kara jizoku kanōna shakai o tsukuru (Building a sustainable society from the regional level)', *Sekai* (World), 703, pp. 126–35.

Morris, Aldon D. and Carol M. Mueller (eds) (1992), *Frontiers in Social Movement Theory*, New Haven: Yale University Press.

Morrison, Denton E. (1973), 'The Environmental Movement: Conflict Dynamics', *Journal of Voluntary Action Research*, 2, pp. 74–85.

Motojima, Kunio and Kōkichi Syōji (eds) (1980), Chiiki kaihatsu to shakai közö: Tomakomai töbu daikibo kögyö kaihatsu o megutte (Regional development and social structure: A study of a large scale industrial development project in the eastern part of Tomakomai), University of Tokyo Press.

Münch, Richard et al. (2001), *Democracy at Work: A Comparative Sociology of Environmental Regulation in the United Kingdom, France, Germany and the United States*, Westport: Praeger.

Nagai, Susumu, Shun'ichi Teranishi and Tadahumi Yokemoto (eds) (2002), *Kankyö saisei: Kawasaki kara kögai chiiki no saisei o kangaeru* (Environmental regeneration: Thinking on the regeneration of polluted regions from Kawasaki), Yūhikaku.

Nagoya Shinkansen Kögai Soshö Bengodan (The legal counsel for the Nagoya bullet train lawsuit) (1996), *Shizukasa o kaese! Monogatari shinkansen kögai soshö* (Give back our quiet! Tales from the lawsuit against the bullet train pollution), Fūbaisha.

Nakajima, Shin'ichiro (1988), 'Bunka daikakumei toshite no Ikata no tatakai (The Ikata fight as a cultural revolution)', *Kuritïku* (Critique), 12, pp. 31–40.

Nakanishi, Junko (1979), *Toshi no saisei to gesuidö* (Sewerage and urban regeneration), Nippon Hyöronsha.

————— (1983), *Gesuidö: Mizu saisei no tetsugaku* (Sewerage: The philosophy of water recycling), Asahi Shimbunsha.

Nakano, Yasuto, Kōji Abe, Yöichi Murase and Michio Umino (1996), 'Shakaiteki jirenma toshite no gomi mondai (Environmental problems as social dilemmas: Structures of factors affecting the availability of cooperation in waste reduction problems)', *Kankyö shakaigaku kenkyü* (Journal of environmental sociology), 2, pp. 123–39.

Nakata, Minoru (1993), *Chiiki kyödö kanri no shakaigaku* (The sociology of local communal management system), Töshindö.

Nakazawa, Hideo (2001), 'Kankyö undö to kankyö seisaku no sanjügo nen: "Kankyö" o teigi suru kökyösei no közö tenkan (Structural transformation of relationship between environmental policies and environmental movements: An overview of thirty-five years)', *Kankyö shakaigaku kenkyü* (Journal of environmental sociology), 7, pp. 85–98.

Nippon Shakaigakkai (The Japan Sociological Society) (2000), *Shakaigaku hyöron* (Japanese sociological review), 50 (4), Tokushü: Nijüichi seiki e no shakaigakuteki sözöryoku: Atarashii kyödösei to kökyösei (Special Edition: Sociological imagination toward the 21st century: New communality and publicness).

Nishihara, Kazuhisa and Manabu Sugimoto (2000), 'Nihon no shakaigaku: 'Shakaigaku hyöron' ni miru riron shakaigaku no gojü nen (Japanese sociology: The fifty-year history of Japanese sociological theory in Japanese Sociological Review)', *Jōkyō* (Situation), 11 (7), Hachigatsu go bessatsu: Gendai shakaigaku no saizensen, 3 (August special edition: The frontier of modern sociology, 3), pp. 305–27.

Nishio (Matsuzawa), Hiroki (2002), 'Seikyō bunri soshō undō ni okeru aidentiï to hō: Minō chūkonhi iken sosyō, Osaka sokui no rei, Daijōsai iken sosyō o jirei toshite (Identity and law in the 'litigation movements for the separation of the state and religious institutions: A comparative analysis of the Minō monument to the war dead, the Osaka enthronement ceremony and the great thanksgiving festival following the enthronement ceremony)', *Hōshakaigaku* (The sociology of law), 56, pp. 234–51.

Nitagai, Kamon (2001), 'Shimin no fukusūsei: Kon'nichi no sei o meguru "shutaisei" to "kōkyōsei"' (The plurality of citizens: 'Identity' and 'publicness' centering on existence), *Chiiki shakaigakkai nenpō 13* (Annals of Region and Community Studies, vol. 13), pp. 38–56.

Obara, Ryōko (1988), 'Genpatsu yori mo inochi ga daiji (Life is more important than nuclear power)', *Kuritïku* (Critique), 12, pp. 21–30.

Obitani, Hiroaki (2000), 'Gyogyōsha ni yoru shokurin undō no tenkai to seikaku henyō: Ryūiki hozen undō kara kankyō shigen sōzō undō e (The development and transformation of fishing people's treeplanting movement: From watershed protection to environment and resource creation)', *Kankyō shakaigaku kenkyū* (Journal of environmental sociology), 6, p. 148–62.

Offe, Claus (1985), 'New Social Movements: Challenging the Boundaries of Institutional Politics', *Social Research,* 52 (4), pp. 817–68.

Okabe, Kazuaki (1996), *Intänetto shimin kakumei: Jōhōka shakai, Amerika hen* (The internet citizens' revolution: A report on the informed society from the United States), Ochanomizu Shobō.

———— (2000), *San Furanshisuko hatsu: Shakai henkaku NPO* (From San Francisco: NPO practices for social reform), Ochanomizu Shobō.

Ōkubo, Noriko (1997), 'Kankyō asesumento ni okeru sanka no genjō to kadai (The current conditions and issues of citizen participation in environmental impact assessment procedure)', Kankyō to kōgai (Research on environmental disruption), 27 (1), pp. 33–8.

Olson, Mancur (1965), *The Logic of Collective Action,* Cambridge: Harvard University Press.

Ōno, Teruyuki and Reiko Harvey-Evans (1992), *Toshi kaihatsu o kangaeru* (Thinking on urban development), Iwanami Shoten.

Osaka Bengoshikai Kankyō ken Kenkyūkai (Osaka Bar Association Environmental Rights Research Group) (eds) (1973), *Kankyōken* (Environmental rights), Nippon Hyōronsha.

Osaka Kūkō Kōgai Sosyō Bengodan (Osaka International Airport Pollution Lawsuit Legal Counsel) (eds) (1986), *Osaka kūkō kōgai saiban kiroku, zen roku kan* (Records of the Osaka International Airport pollution trial, six volumes), Daiichi Hōki Shuppan.

Ōsawa, Shūsuke (1988), *Gendaigata soshō no Nichibei hikaku* (Comparative study of Japanese and American 'contemporary lawsuits'), Kōbundō.

Ōtake, Hideo (1994), *Sengo seiji to seijigaku* (Postwar politics and political science in Japan), University of Tokyo Press.

Ōtsuka, Yoshiki (1998), 'Idenshi kumikae sakumotsu o meguru kankyō mondai to kagaku gijutsu no sōgoteki kōchiku (Interactive construction of environmental problems: A case study of transgenic crops)', *Kankyō shakaigaku kenkyū* (Journal of environmental sociology), 4, pp. 93–106.

Parsons, Talcott (1978), *Action Theory and the Human Condition*, New York: Free Press.

Perrow, Charles (1979), 'The Sixties Observed', in Mayer N. Zald and John D. McCarthy (eds), *The Dynamics of Social Movements*, Cambridge: Winthrop, pp. 192–211.

Pestoff, Victor A. (1998), *Beyond the Market and State: Social Enterprises and Civil Democracy in a Welfare Society*, Aldershot: Ashgate.

Pharr, Susan J. and Robert D. Putnam (eds) (2000), *Disaffected Democracies: What's Troubling the Trilateral Countries?*, Princeton: Princeton University Press.

Polletta, Francesca and James M. Jasper (2001), 'Collective Identity and Social Movements', *Annual Review of Sociology*, 27, pp. 283–305.

Popper, Karl R. (1957), *The Poverty of Historicism*, London: Routledge and Kegan Paul.

Putnam, Robert D. (2000), *Bowling Alone: The Collapse and Revival of American Community*, New York: Simon and Schuster.

Redclift, Michael and Graham Woodgate (eds) (1997), *The International Handbook of Environmental Sociology*, Cheltenham: Edward Elgar.

Rokumoto, Kahei (1991), '"Gendaigata sosyō" to sono kinō ('Contemporary lawsuits' and their function)', *Hōshakaigaku* (Legal sociology), 43, pp. 2–12.

Saikō Saibansyo Daihōtei (Supreme Court Tribunal) (1982), 'Osaka Kokusai Kūkō kōgai sosyō jōkokushin hanketsu (Osaka International Airport pollution Supreme Court decision)', *Hanrei jihō* (Precedent review), 1025, pp. 45–234.

Salamon, Lester M. (1992), *America's Nonprofit Sector*, New York: The Foundation Center.

———— (1995), *Partners in Public Service: Government-Nonprofit Relations in the Modern Welfare State*, Baltimore: Johns Hopkins University Press.

Samuta, Hikaru (2001), 'Ōshū sasuteinaburu shitei no tenkai (The development of European sustainable cities projects)', *Kankyō to kōgai* (Research on environmental disruption), 31 (1), pp. 36–43.

Sasaki, Takeshi and Tae-Chang Kim (eds) (2001–2), *Kōkyō tetsugaku* (Public philosophy), ten volumes, University of Tokyo Press.

Satō, Yoshiyuki (ed.) (1988), *Joseitachi no seikatsu nettwāku: Seikatsu kurabu ni tsudou hitobito* (Womens' network: The people gathering in the Seikatsu club consumer's co-op), Bunshindō.

Sax, Joseph L. (1971), *Defending the Environment: A Strategy for Citizen Action*, New York: Alfred A. Knopf.

Scheer, Hermann (2002), *The Solar Economy: Renewable Energy for a Sustainable Global Future*, London: Earthscan.

Seki, Reiko (2001), 'Kankyōken no shisō to undō: "Teikō suru kankyōken" kara "sanka to jichi no kankyōken" e (Thoughts and movements for environmental rights: From the 'environmental rights of protest' to the 'environmental rights of participation and self-government')', in Kōichi Hasegawa (ed.),

Kōza kankyō shakaigaku, 4: Kankyō undō to seisaku no dainamizumu (Environmental sociology in Japan, vol. 4: Dynamism of environmental movements and policy), Yūhikaku, pp. 211–36.

Sendādo Kurabu (Sendādo Club) (eds) (1991), *Kaettekita Sendādo Mappu* (The Map of Sendai citizen's acitivities, revised edition), Katatsumurisha.

Sendai-Miyagi NPO Sentä (Sendai-Miyagi NPO Center) (eds) (2001), *Sendai-Miyagi NPO sentä no shigoto: NPO sapöto, Sendai moderu* (The work of the Sendai-Miyagi NPO center: NPO support, Sendai model).

Sendai-Miyagi NPO Sentä (Sendai-Miyagi NPO Center) (eds) (1999), *Shimin katsudō handobukku* (The citizens' activities handbook), City of Sendai.

Sendai Shimin Ombuzuman (Sendai Citizens' Ombudsman) (1999), *Kanpeki o tsuku* (Hitting the governmental wall), Mainichi Shimbunsha.

Shindō, Kōji (1983), 'Gendaigata sosyō to sono yakuwari ('Contemporary lawsuits' and its role)', *Kihon hōgaku, 8: Funsō* (Basic legal study, vol. 8: Conflicts), Iwanami Shoten, pp. 305–33.

Shiobara, Tsutomu (ed.) (1989), *Shigen dōin to soshiki senryaku: Undōron no Shimparadaimu* (Resource mobilization and organizational strategies: New paradigms in movement theory), Shin'yōsha.

'Shizen Enerugï Sokushinhō' Suishin Nettowäku (Green Energy Law Network) (eds) (2000), *Nisenjünen shizen enerugï sengen* (Charter for renewable energy by 2010), Nanatsu Mori Shokan.

Shizenken Seminä Hōkokusho Sakusei Iinkai (Environmental Rights Seminar Report Committee) (eds) (1998), *Hōkoku: Nihon ni okeru 'shizen no kenri' undō* (Report: Environmental rights movements in Japan), Shizen no Kenri Seminä (Environmental rights seminar).

Shōji, Kōkichi (1999), *Chikyü shakai to shimin renkei: Gekiseiki no kokusai shakaigaku e* (Global society and citizens' networks: Towards a new international sociology in a drastically changing age), Yūhikaku.

Shōji, Hikaru and Ken'ichi Miyamoto (1964), *Osoru beki kōgai* (Awful industrial pollutions), Iwanami Shoten.

Shrader-Frechette, Kristin Sharon (1981), *Environmental Ethics*, Pacific Grove: Boxwood Press.

Smelser, Neil (1963), *The Theory of Collective Behavior*, New York: Free Press.

Snow, David A., E. Burke Rochford Jr, Steven K. Worden and Robert D. Benford (1986), 'Frame Alignment Processes, Micromobilization and Movement Participation', *American Sociological Review*, 51, pp. 464–81.

Stone, Christopher D. (1972), 'Should Trees Have Standing?: Toward Legal Rights for Natural Objects', *Southern California Law Review*, 45, pp. 450–501.

Suda, Harumi, Mitsuru Tanaka and Kazunori Kumamoto (eds) (1992), *Kankyō jichitai no sōzō* (The creation of local governments for environmental initiative), Gakuyō Shobō.

Sung, Won-Cheol (1998), '"Risuku shakai" no tōrai o tsugeru jümin tōhyō undō: Niigata Ken, Maki Machi to Gifu Ken, Mitake Chō no jirei o tegakari ni (The rise of local referendum movements to herald the advent of a 'risk society': The cases of Maki Town, Niigata Prefecture and Mitake Town, Gifu Prefecture)', *Kankyō shakaigaku kenkyü* (Journal of environmental sociology), 4, pp. 60–75.

Suzuki, Tōru (2002), 'Shimin füsha to gurïn fando (The citizen's communal wind generator and Green Fund)', *Kankyö shakaigaku kenkyü* (Journal of environmental sociology), 8, pp. 74–9.

Takada, Akihiko (1990), 'Kusa no ne shimin undö no nettowäkingu: Musashinoshi no jirei kenkyü o chüshin ni (Grassroots citizens' move-ments' networking: From the case study of Musashino city)', in Shakai Undöron Kenkyükai (Research group on social movement theories) (eds), *Shakai undöron no tögö o mezashite* (Toward the study of social movement theory), Seibundö, pp. 203–46.

———— (1995), 'Kankyö mondai e no sho apuröchi to shakai undöron (Environmental sociology and the theory of social movements)', *Shakaigaku hyöron* (Japanese sociological review), 45 (4), pp. 16–38.

Takagi, Jinzaburö (1991), *Shimo Kita Hantö Rokkasho Mura kakunenryö saikuru shisetsu hihan* (Criticizing the nuclear fuel cycle project at Rokkasho Village, Shimokita Peninsula), Nanatsu Mori Shoten.

———— (1999), *Shimin kagakusha toshite ikiru* (My life as a citizen scientist), Iwanami Shoten.

Takegawa, Shögo (1992), *Chiiki shakai keikaku to jümin seikatsu* (Regional social planning and community life), Chüö University Press.

Takeuchi, Keiji (1998), *Chikyü ondanka no seijigaku* (The politics of global warming), Asahi Shimbunsha.

Takubo (Hirabayashi), Yüko (1996), 'Kariforunia shü "genshiryoku anzenhö" no seiritsu katei (The enactment of the "nuclear moratorium law" in California)', *Kankyö shakaigaku kenkyü* (Journal of environmental sociology), 2, pp. 91–108.

———— (1997), 'Maki Machi "jümin töhyö o jikkö suru kai' no tanjö, hatten to seikö (The emergence, development and success of 'Association for the referendum' in Maki)', *Kankyö shakaigaku kenkyü* (Journal of environmental sociology), 3, pp. 131–48.

Tanaka, Mitsuru (2002), 'Jichitai enerugï seisaku no köchiku ni mukete (Constructing an energy policy on local government level)', *Kankyö shakaigaku kenkyü* (Journal of environmental sociology), 8, pp. 38–53.

Tanaka, Shigeaki (1979), *Saiban o meguru hö to seiji* (Politics and law in judicial trials), Yühikaku.

Tanaka, Shigeru (1997), 'Kasen kankyö jigyö toshite no "tashizengata kawazukuri": 1970 nendai ikö ni okeru kensetsushö, kasen kankyö gyöseishi ('Renaturalization of the river' as a part of environmental river projects: The history of river administration in Japan since the 1970s)', *Kankyö shakaigaku kenkyü* (Journal of environmental sociology), 3, pp. 58–71.

Tanase, Takao (1972), 'Saiban o meguru infuruensu katsudö (Influential campaigns in judicial trials)', in Takeyoshi Kawashima (ed.), *Höshakaigaku köza, 5: Funsö kaiketsu to hö* (Legal sociology in Japan, vol. 5: Conflict resolution and law), Iwanami Shoten, pp. 306–54.

Tarrow, Sidney (1994), *Power in Movement: Social Movements, Collective Action and Politics,* Cambridge: Cambridge University Press.

Terada, Ryöichi (1998a), 'Kankyö NPO (minkan hieiri soshiki) no seidoka to kankyö undö no hen'yö (The institutionalization of environmental NPOs and its impact on environmental movements)', *Kankyö shakaigaku kenkyü* (Journal of environmental sociology), 4, pp. 7–23.

———— (1998b), 'Kankyö undö to kankyö seisaku: Kankyö undö no seidoka to kusanone minshushugi no nichibei hikaku (Environmental movements and environmental policy: Comparative study of institutionalization of environmental movements and grassroots democracy in Japan and the US)', in Harutoshi Funabashi and Nobuko Iijima (eds), *Kōza shakaigaku, 12: Kankyō* (Sociology in Japan, vol. 12: Environment), University of Tokyo Press, pp. 133–62.

———— (2001), 'Chikyū kankyō ishiki to kankyō undō: Chiiki kankyō shugi to chikyū kankyō shugi (Global environmental consciousness and environmental movements: Regional environmentalism and global environmentalism)', in Nobuko Iijima (ed.), *Kōza kankyō shakaigaku, 5: Ajia to sekai: Chiiki shakai kara no shiten* (Environmental sociology in Japan, vol. 5: Asian and global environmental problems: Sociological studies of local origins), Yūhikaku, pp. 233–58.

Teranishi, Shun'ichi (1997), '"Kankyō higai" ron josetsu (Introduction to the theory of 'environmental disruption')', in Takehisa Awaji and Shun'ichi Teranishi (eds), *Kōgai kankyōhō riron no aratana tenkai* (New developments in the theory of pollution and environmental law), Nippon Hyōronsha, pp. 92–104.

Toda, Kiyoshi (1994), *Kankyōteki kōsei o motomete* (In search of environmental justice), Shin'yōsha.

Tokuno, Sadao (2001), 'Nōgyō ni okeru kankyō hakai to kankyō sōzō (Environmental damage and creation in agriculture)', in Hiroyuki Torigoe (ed.), *Kōza kankyō shakaigaku, 3: Shizen kankyō to kankyō bunka* (Environmental sociology in Japan, vol. 3: The natural environment and environmental culture), Yūhikaku, pp. 105–32.

Tominaga, Ken'ichi (1965) *Shakai hendō no riron* (Theory of social change), Iwanami Shoten.

———— (2002), 'Shakai shisutemuron kara mita kankyō jōhōgaku (Environmental information science viewed from social system theory)', *Musashi Kōgyō Daigaku kankyō jōhō gakubu kiyō* (Musashi Institute of Technology, Faculty of Environmental and Information Studies bulletin), 3, pp. 7–26.

Tomizawa, Kenji and Kiyofumi Kawaguchi (eds) (1997), *Hieiri, kyōdō sekutā no riron to genjitsu: Sankagata shakai shisutemu o motomete* (Theory and reality of the nonprofit and cooperative sectors: In search of a social system of citizen participation), Nippon Keizai Hyōronsha.

Torigoe, Hiroyuki (1995), 'Soko ni sumu mono no kenri (The rights of those that live there)', in Kō Mito and Yoshiyuki Satō (eds), *Kankyō hakai: Shakai shokagaku no ōtō* (Environmental disruption: Responses from the social sciences), Bunshindō, pp. 178–98.

———— (1997), *Kankyō shakaigaku no riron to jissen: Seikatsu kankyōshugi no tachiba kara* (Theory and practice of environmental sociology: From the standpoint of life-environmentalism), Yūhikaku.

———— (2001a), 'Kankyō kyōzon eno apurōchi (An approach to environmental coexistence)', in Nobuko Iijima, Hiroyuki Torigoe, Kōichi Hasegawa and Harutoshi Funabashi (eds), *Kōza kankyō shakaigaku, 1: Kankyō shakaigaku no shiten* (Environmental sociology in Japan, vol. 1: Perspectives of environmental sociology), Yūhikaku, pp. 63–87.

———— (2001b), 'Ningen ni totte no shizen: Shizen hogoron no saikentō ('Nature' for humans: Nature preservation re-examined)', in Hiroyuki

Torigoe (ed.), *Köza kankyö shakaigaku, 3: Shizen kankyö to kankyö bunka* (Environmental sociology in Japan, vol. 3: The natural environment and environmental culture), Yühikaku, pp. 1–23.

Torigoe, Hiroyuki (ed.) (1989), *Kankyö mondai no shakai riron* (Social theory of environmental problems), Ochanomizu Shobö.

Torigoe, Hiroyuki and Yukiko Kada (eds) (1984), *Mizu to hito no kankyöshi: Biwako hökokusho* (Environmental history of water and people: The Lake Biwa report), Ochanomizu Shobö.

Touraine, Alain (1985), 'An Introduction to the Study of Social Movements', *Social Research*, 52 (4), pp. 749–87.

Touraine, Alain, Zsuzsa Hegedus, Francois Dubet and Michel Wieviorka (1980), *La Prophétie Anti-Nucléaire*, Paris: Editions du Seuil.

Toyoda, Makoto (1982), 'Kögai saiban to shihö no kinö (Judicial trials and functions of anti-pollution lawsuits)', in Toshitaka Ushiomi, Hirohisa Kitano, Oda Shigemoto and Tadasuke Toryü (eds), *Gendai shihö no kadai* (Issues in contemporary judiciary), Keisö Shobö, pp. 415–49.

Tsukahara, Eiji (1990), 'Saibankan keireki to saiban ködö (The career of judges and their judicial decisions)', *Höritsu jihö* (Law Review), 62 (9), pp. 26–33.

Turner, Ralf H. and Lewis M. Killian (1972), *Collective Behavior*, 2nd ed., Englewood Cliffs, N.J., Prentice-Hall.

Ueta, Kazuhiro, Hitoshi Ochiai, Yoshifusa Kitabatake and Shun'ichi Teranishi (1991), *Kankyö keizaigaku* (Environmental economics), Yühikaku.

Ui, Jun (1974), 'Kögai saiban no imi to yakuwari (The significance and role of anti-pollution judicial trials)', *Kögai genron, hokan ni: Kögai jümin undö* (Principles of pollution, supplement 2: Antipollution residents' movements), Aki Shobö, pp. 143–203.

———— (ed.) (1992), *Industrial Pollution in Japan*, United Nations University Press.

Ukai, Teruyoshi (1992), *Okinawa kyodai kaihatsu no ronri to hihan: Shin Ishigaki kükö kensetsu hantai undö kara* (Logic and problems of large scale development projects in Okinawa: From a study on the movement opposed to construction of the Shin Ishigaki Airport), Shakai Hyöronsha.

Umino, Michio (1991), 'Shakaiteki jirenma kenkyü no shatei (The scope of research into social dilemmas)', in Kazuo Seiyama and Michio Umino (eds), *Chitsujo mondai to shakaiteki jirenma* ('Problems of order' and social dilemmas), Häbesutosha.

———— (2001), 'Gendai shakaigaku to kankyö shakaigaku o tsunagu mono: Sögo köryü no genjö to kanösei (Between current sociological study and environmental sociology: The actualities and possibilities of interaction)', in Nobuko Iijima, Hiroyuki Torigoe, Köichi Hasegawa and Harutoshi Funabashi (eds), *Köza kankyö shakaigaku, 1: Kankyö shakaigaku no shiten* (Environmental sociology in Japan, vol. 1: Perspectives of environmental sociology), Yühikaku, pp. 155–86.

Wakabayashi, Keiko (2001), 'Jinkö mondai to kankyö mondai: Chügoku no jirei o chüshin ni (The population problem and environmental problems: Focusing on the example of China)', in Nobuko Iijima (ed.), *Köza kankyö shakaigaku, 5: Ajia to sekai: Chiiki shakai kara no shiten* (Environmental sociology in Japan, vol. 5: Asian and global environmental problems: Sociological studies of local origins), Yühikaku, pp. 121–52.

Wakita, Ken'ichi (2001), 'Chiiki kankyō mondai o meguru "jōkyō no teigi no zure" to "shakaiteki kontekusuto": Shiga Ken ni okeru sekken undō o motoni ('Social contexts' and 'gaps in the definition of situations' in community environmental problems: From the case of the soap movements in Shiga Prefecture)', in Harutoshi Funabashi (ed.), *Kōza kankyō shakaigaku, 2: Higai to kaiketsu katei* (Environmental sociology in Japan, vol. 2: Environmental perpetration, suffering and the process of solving environmental problems), Yūhikaku, pp. 177–206.

Waseda Daigaku Bungakubu Shakaigaku Kenkyūshitsu (Waseda University, Sociology, School of Literature, Arts and Science) (1988), *Chōsa hōkokusho: Ikego beigun kazoku jūtaku kensetsu mondai o megutte* (Report on the disputes about the construction of the US military housing complex at Ikego forests).

Watanabe, Noboru (1999), 'Chiiki shakai ni okeru ishi kettei shudan toshite no jūmin tōhyō shikō no haikei (A social movement for local referendum to express the will of local people)', *Shakaigaku nenpō* (Annual reports of the Tohoku Sociological Society), 28, pp. 1–30.

Yamaguchi, Yasushi, Harukichi Satō, Shigeki Nakajima and Motoaki Ozeki (eds) (2003), *Atarashii Kōkyōsei* (Frontiers of Publicness), Yūhikaku.

Yamamura, Tsunetoshi (ed.) (1998), *Kankyō NGO: Sono katsudō, rinen to kadai* (Environmental NGOs: Their activities, ideologies and issues), Shinzansha.

Yamamuro, Atsushi (1998), 'Genshiryoku hatsudenjo kensetsu mondai ni okeru jūmin no ishi hyōji: Niigata Ken, Maki Machi o jirei ni (Expressing the opinion on a controversial issue: The case of nuclear power plant issue in Maki Town, Niigata Prefecture)', *Kankyō shakaigaku kenkyū* (Journal of environmental sociology), 4, pp. 188–203.

Yamaoka, Yoshinori (2001), 'Hieiri, kōeki hōjin seido o torimaku atarashī ugoki (New actions related to the non-profit, public interest, corporate system)', in Nakamura Yōichi and Nihon NPO Sentā (Japan NPO Center) (eds), *Nihon no NPO, 2001* (Japanese NPOs 2001), Nippon Hyōronsha, pp. 2–26.

Yonemoto, Shōhei (1994), *Chikyū kankyō mondai towa nanika* (What are global environmental problems?), Iwanami Shoten.

Yoshida, Fumikazu (1989), *Haiteku osen* (Hi-tech pollution), Iwanami Shoten.

———— (2002), *The Economics of Waste and Pollution Management in Japan*, Tokyo: Springer.

Yoshikane, Hideo (1996), 'Fīrudo kara manabu kankyō bunka no jūyōsei (The importance of environmental culture found through fieldwork)', *Kankyō shakaigaku kenkyū* (Journal of environmental sociology), 2, pp. 38–49.

Yoshihara, Naoki (2000), 'Chiiki jūmin soshiki ni okeru kyōdōsei to kōkyōsei (Communality and publicness in neighborhood groups: The case of the chonaikai)', *Shakaigaku hyōron* (Japanese sociological review), 50 (4), pp. 140–53.

———— (2002), *Toshi to modernity no riron* (The city and modernity), University of Tokyo Press.

Yoshimoto, Tetsuro (1995), *Watashi no jimoto gaku: Minamata kara no hasshin* (On-site community studies: Calling from Minamata), NEC Kurieitibu.

Yoshimura, Isao (1984), *Gomi to toshi seikatsu: Kankyō asesumento o megutte* (Garbage and urban life: The role of environmental impact assessment), Iwanami Shoten.

Yukawa, Jirö (1990), 'Gyösei jiken ni okeru saibankan kaidö, kyögikai (On the role of the meeting of judges for the guidance of the application of administrative laws)', *Höritsu jihö* (Law review), 62 (9), pp. 34–46.

Yüki, Emi (1988), *Sayonara genpatsu gaidobukku* (A guidebook for goodbye to nuclear energy), Seiyüsha.

Zald, Mayer N. and John D. McCarthy (1987), *Social Movements in an Organizational Society: Collected Essays*, New Brunswick: Transaction.

Name index

Subject Index

www.ingramcontent.com/pod-product-compliance
Lightning Source LLC
Chambersburg PA
CBHW070556270326
41926CB00013B/2332